Forum for Interdisciplinary M

Editors-in-Chief

Viswanath Ramakrishna, University of Texas, Richardson, USA

Zhonghai Ding, University of Nevada, Las Vegas, USA

Editorial Board

Ashis Sengupta, Indian Statistical Institute, Kolkata, India

Balasubramaniam Jayaram, Indian Institute of Technology Hyderabad, Hyderabad, India

P. V. Subrahmanyam, Indian Institute of Technology Madras, Chennai, India

Ravindra B. Bapat, Indian Statistical Institute, New Delhi, India

The *Forum for Interdisciplinary Mathematics* is a Scopus-indexed book series. It publishes high-quality textbooks, monographs, contributed volumes and lecture notes in mathematics and interdisciplinary areas where mathematics plays a fundamental role, such as statistics, operations research, computer science, financial mathematics, industrial mathematics, and bio-mathematics. It reflects the increasing demand of researchers working at the interface between mathematics and other scientific disciplines.

More information about this series at https://link.springer.com/bookseries/13386

Guruprasad Samanta

Deterministic, Stochastic and Thermodynamic Modelling of some Interacting Species

Guruprasad Samanta
Department of Mathematics
Indian Institute of Engineering Science
and Technology, Shibpur
Howrah, West Bengal, India

ISSN 2364-6748 ISSN 2364-6756 (electronic)
Forum for Interdisciplinary Mathematics
ISBN 978-981-16-6314-7 ISBN 978-981-16-6312-3 (eBook)
https://doi.org/10.1007/978-981-16-6312-3

© Springer Nature Singapore Pte Ltd. 2021
This work is subject to copyright. All rights are reserved by the Publisher, whether the whole or part of the material is concerned, specifically the rights of translation, reprinting, reuse of illustrations, recitation, broadcasting, reproduction on microfilms or in any other physical way, and transmission or information storage and retrieval, electronic adaptation, computer software, or by similar or dissimilar methodology now known or hereafter developed.
The use of general descriptive names, registered names, trademarks, service marks, etc. in this publication does not imply, even in the absence of a specific statement, that such names are exempt from the relevant protective laws and regulations and therefore free for general use.
The publisher, the authors and the editors are safe to assume that the advice and information in this book are believed to be true and accurate at the date of publication. Neither the publisher nor the authors or the editors give a warranty, expressed or implied, with respect to the material contained herein or for any errors or omissions that may have been made. The publisher remains neutral with regard to jurisdictional claims in published maps and institutional affiliations.

This Springer imprint is published by the registered company Springer Nature Singapore Pte Ltd.
The registered company address is: 152 Beach Road, #21-01/04 Gateway East, Singapore 189721, Singapore

Dedicated to the memory of my respected teacher and Ph.D. supervisor

Late Prof. Charu Gopal Chakrabarti

Preface

In theoretical ecology and applied mathematics, the logistic and Lotka--Volterra models have long been considered as seminal examples of modelling and dynamics to describe various dynamical aspect of population interactions and will be the most fundamental models in population biology in the future. Population dynamics is an important subject in mathematical ecology. A central problem is to study the long-term behaviour of modelling systems. Most of these systems are governed by various evolutionary equations. It was understood how different forms of regulatory mechanisms, like birth and death, competition, consumption and the like, result in changes of the stability and dynamics of ecological systems. The present book brings this understanding to the attention of students and researchers. It does so with a deep and unique insight into the mathematical richness of basic ecological models. The topics of this book are as follows:

- In Chap. 1, we have discussed the dynamical behaviour of the basic mathematical models of single and two interacting prey and predator species devised by Lotka and independently by Volterra around 1925. These are the starting models of mathematical ecology. Here, we have also discussed several possible modifications of this model.
- The effect of toxicants on ecological systems is an important issue from mathematical and experimental points of view. Chapter 2 deals with the dynamical models of single-species system in a polluted environment.
- As we know, interactive populations often live in a fluctuating environment. For example, physical environmental conditions such as temperature and humidity and the availability of food, water and other resources usually vary in time with seasonal or daily variations. Therefore, more realistic models should be nonautonomous systems. In Chap. 3, we have discussed the dynamical behaviour of different nonautonomous two species systems in a polluted environment.
- Chapter 4 aims to study the influence of environmental noise in Gompertzian and Logistic growth models. The stability behaviour of these models in a deterministic environment and the corresponding model in a stochastic environment have been analysed here.

- In recent years, scientists have become increasingly aware of the fact that most natural phenomena do not follow strictly deterministic laws but rather oscillate randomly about some average behaviour. This is especially true in the ecological models where environmental influences should be taken as stochastic. Chapter 5 analyses stability behaviour in randomly fluctuating versus deterministic environments of two interacting species.
- Chapter 6 deals with stochastic analysis of a demographic model of urbanization.
- Chapter 7 develops a non-equilibrium thermodynamic model of interacting species.
- In Chap. 8, we have studied the stability behaviour of a social group by means of loop analysis, thermodynamic criteria of stability and stochastic criteria of stability.

I expect that the work reported in this book will generate much more interests among the students and researchers of mathematical biological systems. Further, this book can be used for a text for postgraduate courses in applied mathematics programmes.

It is my pleasure to express my grateful thanks to Dr. Swarnali Sharma of the Department of Biological Sciences, University of Notre Dame, USA; Dr. Ricardo Gómez Aíza of the Institute of Mathematics, National Autonomous University of Mexico, Mexico; Dr. Alakes Maiti of the Department of Mathematics, Vidyasagar Evening College, Kolkata, India; and Dr. Debasis Manna of the Department of Mathematics, Surendranath Evening College, Kolkata, India, Dr. Prosenjit Sen of the Department of Mathematics, Om Dayal Group of Institutions, Howrah, India; and my beloved students Debgopal Sahoo, Bijoy Kumar Das, Nirapada Santra, who offered their immense help at various stages of the preparation of this book. Finally, I wish to express my special thanks to Mr. Shamim Ahmad, Senior Editor, Mathematical Sciences, Springer, and the staff of Springer for their help and cooperation.

Howrah, India Guruprasad Samanta

Contents

About the Author

Guruprasad Samanta is a Professor, Higher Administrative Grade, at the Department of Mathematics, Indian Institute of Engineering Science and Technology, Shibpur, Howrah, India. He earned his Ph.D. degree in applied mathematics from the University of Calcutta, India, in 1991. He is a Premchand Roychand Scholar of the University of Calcutta and awarded Mouat Medal at the convocation of the University of Calcutta in 1996. His research areas are mathematical biology and operations research. He has published 9 books and more than 160 papers in reputed journals. He earned his D.Sc. degree in mathematics from the SKM University, Dumka, India, in 2020.

Chapter 1
Dynamical Models of Single and Predator–Prey Species

1.1 Introduction

In this chapter, we have discussed the dynamical behaviour of the basic mathematical models of single and two interacting prey and predator species devised by Lotka and independently by Volterra around 1925. These are the starting models of mathematical ecology. Lotka–Volterra model has two equilibrium (or steady state) positions; one is trivial equilibrium position, and another is non-trivial equilibrium position. The trivial equilibrium position is always unstable, and the non-trivial equilibrium position is stable but not asymptotically; this is so because this model has no internal mechanism to stabilize (asymptotically) the non-trivial equilibrium position. Next, we have discussed the effects of intraspecific competition among the prey population of this model. Here, we have also discussed several possible modifications of this model.

1.2 Malthus Population Growth Model

One of the first researchers into population dynamics was British scientist Thomas Robert Malthus (1766–1834). He proposed a mathematical model of population growth in 1798 based on the idea that 'the rate at which a population grows is directly proportional to its current size". His model, though simple, has become a basis for most future modelling of biological population growth.

Malthus model has been described below in terms of continuous time variable:

$$\frac{1}{N}\frac{\mathrm{d}N}{\mathrm{d}t} = b - d = r \tag{1.1}$$

where $N = N(t)$ denotes the population density or population size of a single species at time t; $b > 0, d > 0$ represents, respectively, *per capita or average* birth rate, *per*

© Springer Nature Singapore Pte Ltd. 2021

G. Samanta, *Deterministic, Stochastic and Thermodynamic Modelling of some Interacting Species*, Forum for Interdisciplinary Mathematics,
https://doi.org/10.1007/978-981-16-6312-3_1

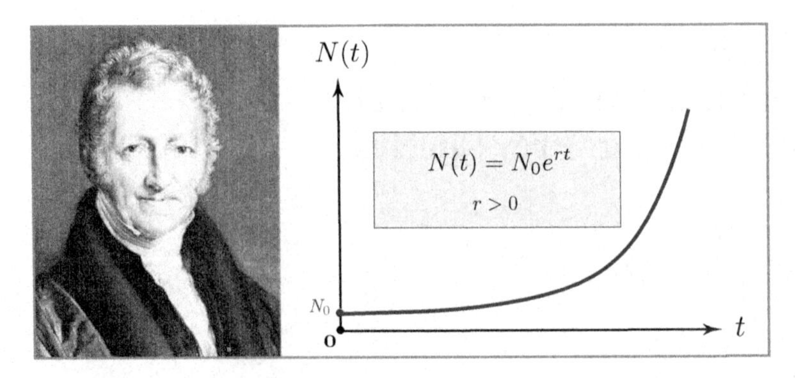

Fig. 1.1 Thomas Robert Malthus (1766–1834)

capita or average death rate and $r \in \mathbb{R}$ denotes *per capita* net population growth rate. From (1.1), we get

$$\int_{t=0}^{t} \frac{dN}{N} = \int_{t=0}^{t} r dt \implies N(t) = N_0 e^{rt}, \tag{1.2}$$

where $N_0 = N(0)$ denotes initial population density or population size at time $t = 0$. Malthus growth is often called *"exponential growth"* (see Fig. 1.1).

Observation 1.1 1. If the *per capita* birth rate is greater than the *per capita* death rate, i.e. $b > d$, then $r > 0 \Rightarrow N(t) \to \infty$ as $t \to \infty \implies$ population size will grow unbounded as time goes.
2. If the *per capita* birth rate is equal to the *per capita* death rate, i.e. $b = d$, then $r = 0 \Rightarrow N(t) = N(0)$ for all time $\implies$ no change in the population size.
3. If the *per capita* birth rate is less than the *per capita* death rate, i.e. $b < d$, then $r < 0 \Rightarrow N(t) \to 0$ as $t \to \infty \implies$ population size will approach to zero as time goes (i.e. extinction).

The cases 2 and 3 are realistic, but the case 1 is not so as the population increases it will demand more food, water, etc., that will not be available in unlimited quantities.

1.3 Logistic Population Growth Model

In 1838, the Belgian mathematician Pierre Francois Verhulst (1804–1849) modified the case 1 of Malthus model (1.1) by introducing an inhibiting term on the right side of (1.1) which is proportional to the population size ($N(t)$) to take into account the competition for resources among members of the population because as the population increases it will demand more food, water, etc., that will not be available in unlimited quantities.

The Verhulst model has been described below in terms of continuous time variable:

$$\frac{1}{N}\frac{\mathrm{d}N}{\mathrm{d}t} = r - cN \quad (r > 0, \quad c > 0), \tag{1.3}$$

where $N = N(t)$ denotes the population density or population size of a single species at time t and the term cN stands for *intraspecific competition* ($c > 0$ is a constant).

As with the Malthus model, this model includes a *per capita* growth rate $r > 0$ (as in case 1) which represents the net *per capita* rate at which the population would grow in the absence of intraspecific (i.e. $c = 0$). The Verhulst model is also called "Logistic population growth model".

From (1.3) separating the variables and integrating, we get (after some simplifications):

$$N(t) = \frac{N_0 K e^{rt}}{N_0 e^{rt} + K - N_0} = \frac{K}{1 + \left(\dfrac{K}{N_0} - 1\right) e^{-rt}} \Rightarrow N(t) \to K \text{ as } t \to \infty,$$

where $K = \dfrac{r}{c}$, provided $r, c > 0$.

The logistic equation and its solution occur in many different fields. The logistic function (i.e. $N(t)$) is also known as the *sigmoid function*, and its graph is known as the *"S-curve"* (see Fig. 1.2). The constant K is known as the *carrying capacity* of the habitat for the species. From a biological point of view, the missing feature of the Malthus model is the concept of carrying capacity. As population increases in size, the environmental ability to support the population decreases and so *per capita* food availability decreases, waste products may accumulate, and birth rates tend to decline while death rates tend to increase. Carrying capacity is the population level at which birth and death rates of a species precisely match, resulting in a stable population size over time.

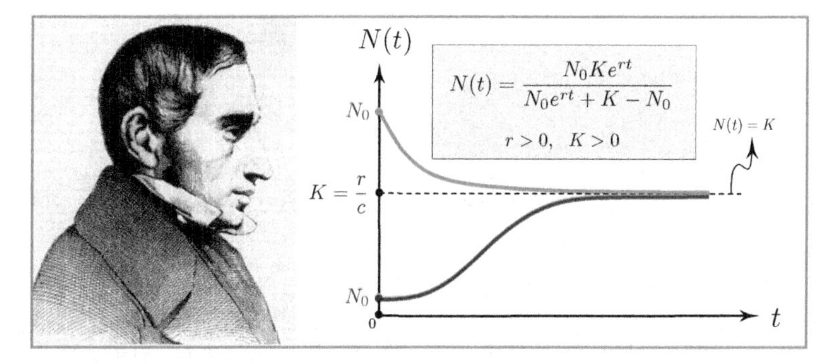

Fig. 1.2 Pierre Francois Verhulst (1804–1849) modified the case 1 of Malthus model (1.1) in 1838 by introducing an inhibiting term

The logistic population growth model (1.3) can also be rewritten as

$$\frac{\mathrm{d}N}{\mathrm{d}t} = rN\left(1 - \frac{N}{K}\right) \quad (r > 0, \ \ K > 0). \tag{1.4}$$

The logistic law explains very well the growth of a bacterial colony in a nutrient medium. This law has also been successfully used to fit data for human populations. The following assumptions have been made for the logistic population growth model:

- The population is uniformly spread throughout the habitat. Thus, the population density (i.e. the population size per unit area) does not vary with position, but it does vary with time.
- There is no immigration or emigration. The population size changes only due to birth and death.

It is observed that

- $N(t)$ cannot cross the straight line $N = K$.
- $N_0 = K = \frac{r}{c} \Rightarrow N(t) = N_0 = K, \ \forall t$.
- Biologically, of course, negative populations do not make sense.

Definition 1 Consider the following first-order autonomous ordinary differential equation:

$$\frac{\mathrm{d}N}{\mathrm{d}t} = f(N). \tag{1.5}$$

1. A constant solution $N(t) = C$ of (1.5) is called an *equilibrium point* (or *stationary point*). An equilibrium point corresponds to $\frac{\mathrm{d}N}{\mathrm{d}t} = f(N) = 0$.
2. An equilibrium point C is called *stable* if $N(t) \to C$ as $t \to \infty$, for all initial conditions $N(0)$.
3. If an equilibrium point is not stable, then it is called *unstable*.

The system (1.3) or (1.5) has equilibrium points:

- $N = 0$, unstable equilibrium point.
- $N = K = \dfrac{r}{c}$, stable equilibrium point.

1.4 Lotka–Volterra Model of Predator–Prey System

Populations do not live in isolation. Everybody has a few enemies here and there. One of the first models to incorporate interactions between predator and prey was proposed in 1925 by the American biophysicist Alfred Lotka and independently by the Italian mathematician Vito Volterra. The Lotka–Volterra model describes interactions between two species in an ecosystem: one *prey* (e.g. rabbits) and one *predator* (e.g. foxes) and assumes the following:

1. Rabbits have unlimited food supply, and so if there were no foxes, the density of rabbits $N_1 = N_1(t)$ will grow exponentially (Malthus law), that is $\frac{dN_1}{dt} = r_1 N_1$, for $r_1 > 0$.
2. In the presence of foxes, $N_1(t)$ is decreasing because of the consumption by foxes at a rate proportional to $N_1 N_2$, where $N_2 = N_2(t)$ denotes the density of foxes at time t. Therefore, $\frac{dN_1}{dt} = r_1 N_1 - b_1 N_1 N_2$, where $r_1 > 0$, $b_1 > 0$ are constants. Here, $b_1 N_1$ represents the density of prey species consumed by one predator in unit time and is known as the *predator functional response* (or the *Volterra functional response* or the *Holling type I response*) on prey species.
3. In the absence of rabbits, the density of foxes $N_2(t)$ will be decreasing exponentially to zero (extinction), i.e. $\frac{dN_2}{dt} = -r_2 N_2$, where $r_2 > 0$ is a constant because if the foxes were left with no prey (rabbits) to eat they would die faster than they could reproduce and would experience exponential population decline.
4. In the presence of rabbits, the density of foxes $N_2(t)$ is increasing at a rate proportional to $N_1 N_2$, that is proportional to the number of encounters between foxes and rabbits. Together with 3, we get $\frac{dN_2}{dt} = -r_2 N_2 + b_2 N_1 N_2$, where $r_2 > 0$ and $b_2 > 0$ are constants. Here, $\frac{b_2}{b_1}$ is called the *conversion factor* and obviously $0 < \frac{b_2}{b_1} < 1 \Rightarrow 0 < b_2 < b_1$.

Thus, we get the following Lotka–Volterra predator–prey model:

$$
\begin{aligned}
\frac{dN_1}{dt} &= r_1 N_1 - b_1 N_1 N_2 = b_1 N_1 \left(\frac{r_1}{b_1} - N_2 \right) \\
\frac{dN_2}{dt} &= -r_2 N_2 + b_2 N_1 N_2 = b_2 N_2 \left(-\frac{r_2}{b_2} + N_1 \right)
\end{aligned}
\tag{1.6}
$$

This is a pair of quadratic differential equations; a completely analytical solution of this system is not possible. But for every value of the parameters r_1, r_2, b_1, b_2 and every initial conditions $N_1(0) = N_{10}$, $N_2(0) = N_{20}$, these can be integrated numerically to get approximate solution.

The Lotka–Volterra model is also a feedback model because the prey population has a positive effect on the size of the predator population, whereas the latter has a negative (inhibiting) effect on the size of the prey population.

1.4.1 Trajectories

From (1.6), we get

$$\frac{dN_2}{dN_1} = \frac{N_2(b_2N_1 - r_2)}{N_1(r_1 - b_1N_2)}. \tag{1.7}$$

Separating the variables and integrating, we get

$$\int_{t=0}^{t} \frac{1}{N_2}(r_1 - b_1N_2)dN_2 = \int_{t=0}^{t} \frac{1}{N_1}(b_2N_1 - r_2)dN_1.$$

$$\Rightarrow r_1 \log \frac{N_2}{N_{20}} - b_1(N_2 - N_{20}) = b_2(N_1 - N_{10}) - r_2 \log \frac{N_1}{N_{10}}$$

$$\Rightarrow \frac{N_2^{r_1} N_1^{r_2}}{N_{20}^{r_1} N_{10}^{r_2}} = e^{b_1(N_2 - N_{20})} e^{b_2(N_1 - N_{10})},$$

where $N_1(0) = N_{10}$ and $N_2(0) = N_{20}$ are initial population sizes of prey and predator species, respectively.

Through each point $P_0 \equiv (N_{10}, N_{20})$ of the positive quadrant, we get a curve of the family

$$e^{-b_2N_1} N_1^{r_2} = \Phi e^{b_1N_2} N_2^{-r_1}, \text{ where } \Phi = \frac{N_{20}^{r_1} N_{10}^{r_2}}{e^{b_1N_{20} + b_2N_{10}}} \text{ is a constant,} \tag{1.8}$$

These curves in the N_1N_2 plane are called the *trajectories* for the Lotka–Volterra model (1.6) (see Fig. 1.3). The N_1N_2 plane is called the *phase plane*.

To investigate the nature of these trajectories, let us first find their points of intersection with lines parallel to the coordinate axes. If we consider the points of intersection with the line $N_2 = c$, we get

$$e^{-b_2N_1} N_1^{r_2} = \Phi e^{b_1 c} c^{-r_1} = \frac{1}{\lambda} \text{ (say)} \Rightarrow F(N_1) = e^{b_2N_1} - \lambda N_1^{r_2} = 0. \tag{1.9}$$

Since $F(0) > 0,\ F(\infty) > 0$, (1.9) either it does not give any positive root or it gives an even number of positive values of N_1. Moreover, the roots of (1.9) are determined by the abscissae of the points of intersection of the following curves:

$$y = e^{b_2 x},\ y = \lambda x^{r_2}. \tag{1.10}$$

It can be easily seen that these two curves intersect in two distinct real points (say, when $\lambda = \lambda_2$) or in two coincident real points (say, when $\lambda = \lambda^\star$) or do not intersect

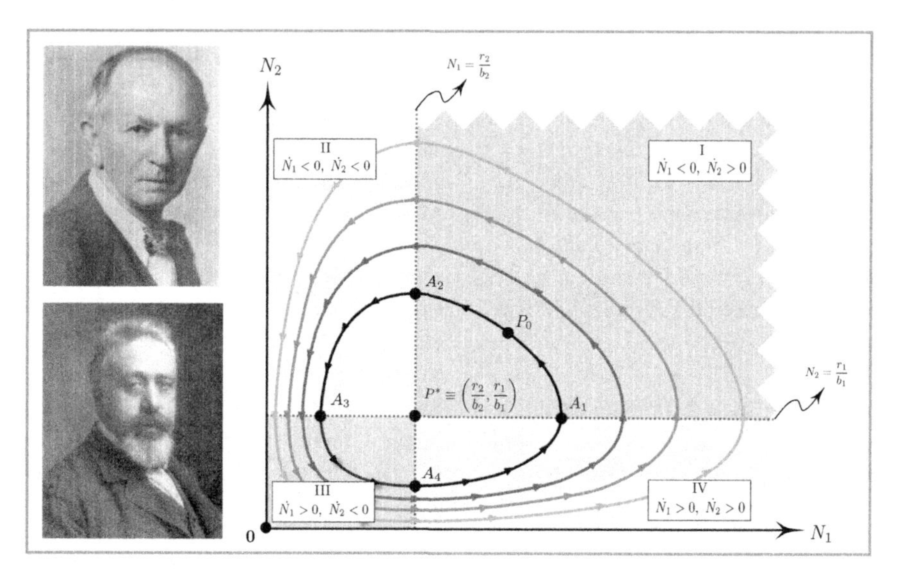

Fig. 1.3 Alfred Lotka (1880–1949, shown above) and Vito Volterra (1860–1940, shown below)

in any real point (say, when $\lambda = \lambda_1$). The condition that these two curves touch each other is obtained by eliminating x between the equations:

$$\lambda x^{r_2} = e^{b_2 x}, \ \lambda r_2 x^{r_2 - 1} = b_2 e^{b_2 x}, \tag{1.11}$$

with the critical value of λ and the value of x as

$$\lambda^\star = \frac{e^{r_2}}{(\frac{r_2}{b_2})^{r_2}}, \ x^\star = \frac{r_2}{b_2}. \tag{1.12}$$

If $\lambda > \lambda^\star$, (1.9) has two real roots and, if $\lambda < \lambda^\star$, it has no real root. Therefore, we conclude that every straight line parallel to the N_1-axis cuts each trajectory in two real (coincident or distinct) points or does not cut in any point. Similarly, every straight line parallel to the N_2-axis cuts each trajectory in two real (coincident or distinct) points or does not cut in any point.

We get some idea of the shapes of the trajectories by considering the signs of the derivatives $\dot{N}_1 = \frac{dN_1}{dt}$ and $\dot{N}_2 = \frac{dN_2}{dt}$ in the four regions I, II, III and IV in which the first quadrant is divided by the lines $N_1 = \frac{r_2}{b_2}$ and $N_2 = \frac{r_1}{b_1}$. From (1.6), we get the signs of $\dot{N}_1, \dot{N}_2$ in these four regions (see Fig. 1.3).

Let the initial point $P_0 \equiv (N_{10}, N_{20})$ be in the region I. In this region, N_1 decreases and N_2 increases, so the point moves in the anticlockwise direction as time goes till it reaches point A_2, where $\dot{N}_2 = 0 \Rightarrow \frac{dN_2}{dN_1} = 0 \Rightarrow$ tangent at A_2 is parallel to N_1-axis. In region II, both N_1 and N_2 decrease and the point continues to move in the anticlockwise direction till it reaches point A_3 , where $\dot{N}_1 = 0 \Rightarrow \frac{dN_2}{dN_1}$ is undefined $\Rightarrow$

tangent at A_3 is parallel to N_2-axis. Similarly, in region III, N_1 increases and N_2 decreases till A_4, and in region IV, both N_1 and N_2 increases till A_1.

Thus, each trajectory is a closed oval curve as shown in Fig. 1.3. It also appears from (1.12) that for every trajectory, points A_2 and A_4, where the line parallel to the N_1-axis touches the trajectory, lie on the line $N_1 = \frac{r_2}{b_2}$. Similarly, the points A_1 and A_3, where the line parallel to the N_2-axis touches the trajectory, lie on the line $N_2 = \frac{r_1}{b_1}$. All the trajectories are thus described in the anticlockwise sense, and they appear to be cramped near the axes since they can only approach but not cross them. It confirms that all solutions of (6) will be periodic closed orbits around the point $P^* \equiv \left(\frac{r_2}{b_2}, \frac{r_1}{b_1}\right)$.

Observation 1.2 Changing the birth and death rates does nothing but changes the period of the oscillation; i.e., no population can dominate, and there is no possibility of either population being driven to extinction. So, prey and predator species coexist together.

Definition 2 Consider the following system of autonomous ordinary differential equations:

$$\begin{aligned} \frac{dN_1}{dt} &= f_1(N_1, N_2) \\ \frac{dN_2}{dt} &= f_2(N_1, N_2) \end{aligned} \quad . \tag{1.13}$$

1. An *equilibrium point* occurs when both $N_1(t)$ and $N_2(t)$ are constants, so $\frac{dN_1}{dt} = 0$ and $\frac{dN_2}{dt} = 0$. In equilibrium point the system is at *rest*. Equilibrium points are also called *stationary points*,or *singular points*,or *critical points*,or *rest points*.
2. An equilibrium point (N_1^*, N_2^*) is called *locally asymptotically stable* if the response to a small perturbation (or, disturbance) approaches zero as time approaches infinity. More formally, this means that

$$\lim_{t\to\infty} N_1(t) = N_1^* \quad \text{and} \quad \lim_{t\to\infty} N_2(t) = N_2^*$$

   for $(N_1(t), N_2(t))$ starting from initial values $P_0 \equiv (N_1(0), N_2(0)) \equiv (N_{10}, N_{20})$ that are sufficiently close to the equilibrium point $\left(N_1^*, N_2^*\right)$. A locally asymptotically stable equilibrium point is also called a *sink* or an *attractor*. The set of all initial values (N_{10}, N_{20}) from which trajectories converge to an attractor is called *domain of attraction*.
3. An equilibrium point $\left(N_1^*, N_2^*\right)$ is called *stable*, but not locally asymptotically stable, if the response to a small perturbation remains small but does not approach zero as time approaches infinity. In this case, the equilibrium point is called a *center*. Here, the solution will return to the initial condition in a periodic fashion; its trajectory is a closed curve around the equilibrium point, often called a *cycle*.

4. An equilibrium point is called *unstable* if initial conditions (near the equilibrium point) produce solutions that repel away from the equilibrium point over time. Near unstable equilibrium points, small changes in initial conditions can drastically affect solutions. An unstable equilibrium point is also called a *source*, or a *repeller*.

Observation 1.3 From the previous discussions, it is clear that the equilibrium point $\mathbf{0} \equiv (0, 0)$ of system (1.6) is unstable and the equilibrium point $P^* \equiv \left(\frac{r_2}{b_2}, \frac{r_1}{b_1}\right)$. is stable (but not asymptotically) which is a centre.

1.4.2 Secular (or Characteristic) Equation for Determining Local Stability

Suppose the system of ODE (1.13) has an equilibrium point at $(\overline{N}_1, \overline{N}_2)$, therefore

$$f_1(\overline{N}_1, \overline{N}_2) = 0 \quad \text{y} \quad f_2(\overline{N}_1, \overline{N}_2) = 0. \tag{1.14}$$

To discuss the linear stability of the model (1.13) at $(\overline{N}_1, \overline{N}_2)$, substitute $N_1(t) = \overline{N}_1 + u_1(t)$, $N_2(t) = \overline{N}_2 + u_2(t)$ in (1.13), where $0 < u_i(t) \ll 1$ (for $i = 1, 2$), as perturbation (or disturbance) is very small. Using Taylor series expansion, we get

$$\begin{aligned}\frac{du_1}{dt} &= f_1\left(\overline{N}_1 + u_1, \overline{N}_2 + u_2\right)\\ &= f_1\left(\overline{N}_1, \overline{N}_2\right) + \left(u_1\frac{\partial}{\partial N_1} + u_2\frac{\partial}{\partial N_2}\right) f_1(N_1, N_2)|_{(\overline{N}_1, \overline{N}_2)}\\ &\quad + \frac{1}{2!}\left(u_1\frac{\partial}{\partial N_1} + u_2\frac{\partial}{\partial N_2}\right)^2 f_1(N_1, N_2)|_{(\overline{N}_1, \overline{N}_2)} + \ldots\end{aligned}$$

and

$$\begin{aligned}\frac{du_2}{dt} &= f_2\left(\overline{N}_1 + u_1, \overline{N}_2 + u_2\right)\\ &= f_2\left(\overline{N}_1, \overline{N}_2\right) + \left(u_1\frac{\partial}{\partial N_1} + u_2\frac{\partial}{\partial N_2}\right) f_2(N_1, N_2)|_{(\overline{N}_1, \overline{N}_2)}\\ &\quad + \frac{1}{2!}\left(u_1\frac{\partial}{\partial N_1} + u_2\frac{\partial}{\partial N_2}\right)^2 f_2(N_1, N_2)|_{(\overline{N}_1, \overline{N}_2)} + \ldots.\end{aligned}$$

Using (1.14) and linearizing (since $u_1(t),\ u_2(t)$ are very small), we obtain

$$\frac{du_1}{dt} = u_1 \frac{\partial f_1}{\partial \overline{N}_1} + u_2 \frac{\partial f_1}{\partial \overline{N}_2} \quad \text{and} \quad \frac{du_2}{dt} = u_1 \frac{\partial f_2}{\partial \overline{N}_1} + u_2 \frac{\partial f_2}{\partial \overline{N}_2} \tag{1.15}$$

where $\frac{\partial f_i}{\partial \overline{N}_j}$ (for $i, j = 1, 2$) denote the values of $\frac{\partial f_i}{\partial N_j}$ at the equilibrium point $(\overline{N}_1, \overline{N}_2)$. Take the solution in the form:

$$u_1 = A_1 e^{\lambda t} \quad \text{and} \quad u_2 = A_2 e^{\lambda t}. \tag{1.16}$$

It is noted that $u_1,\ u_2$ cannot be zero simultaneously because we have really given a perturbation. Substituting $u_1,\ u_2$ from (1.16) in (1.15) and after some simplifications, we get

$$\left(\lambda - \frac{\partial f_1}{\partial \overline{N}_1}\right) A_1 - \frac{\partial f_1}{\partial \overline{N}_2} A_2 = 0$$

$$-\frac{\partial f_2}{\partial \overline{N}_1} A_1 + \left(\lambda - \frac{\partial f_2}{\partial \overline{N}_2}\right) A_2 = 0,$$

In matrix form:

$$\begin{pmatrix} \lambda - \frac{\partial f_1}{\partial \overline{N}_1} & -\frac{\partial f_1}{\partial \overline{N}_2} \\ -\frac{\partial f_2}{\partial \overline{N}_1} & \lambda - \frac{\partial f_2}{\partial \overline{N}_2} \end{pmatrix} \begin{pmatrix} A_1 \\ A_2 \end{pmatrix} = \begin{pmatrix} 0 \\ 0 \end{pmatrix}.$$

Since u_1, u_2 cannot both be zero simultaneously, so A_1, A_2 cannot both be zero simultaneously, and hence,

$$\begin{vmatrix} \lambda - \frac{\partial f_1}{\partial \overline{N}_1} & -\frac{\partial f_1}{\partial \overline{N}_2} \\ -\frac{\partial f_2}{\partial \overline{N}_1} & \lambda - \frac{\partial f_2}{\partial \overline{N}_2} \end{vmatrix} = 0$$

$$\Rightarrow \lambda^2 - \left(\frac{\partial f_1}{\partial \overline{N}_1} + \frac{\partial f_2}{\partial \overline{N}_2}\right)\lambda + \left(\frac{\partial f_1}{\partial \overline{N}_1}\frac{\partial f_2}{\partial \overline{N}_2} - \frac{\partial f_1}{\partial \overline{N}_2}\frac{\partial f_2}{\partial \overline{N}_1}\right) = 0. \tag{1.17}$$

This is known as the *secular* (or *characteristic*) equation. It is a quadratic equation in λ and so it has two solutions λ_1, λ_2. Therefore, the general solution can be written in the form:

$$\begin{aligned} u_1(t) &= a_1 e^{\lambda_1 t} + a_2 e^{\lambda_2 t} \\ u_2(t) &= a_3 e^{\lambda_1 t} + a_4 e^{\lambda_2 t} \end{aligned} \tag{1.18}$$

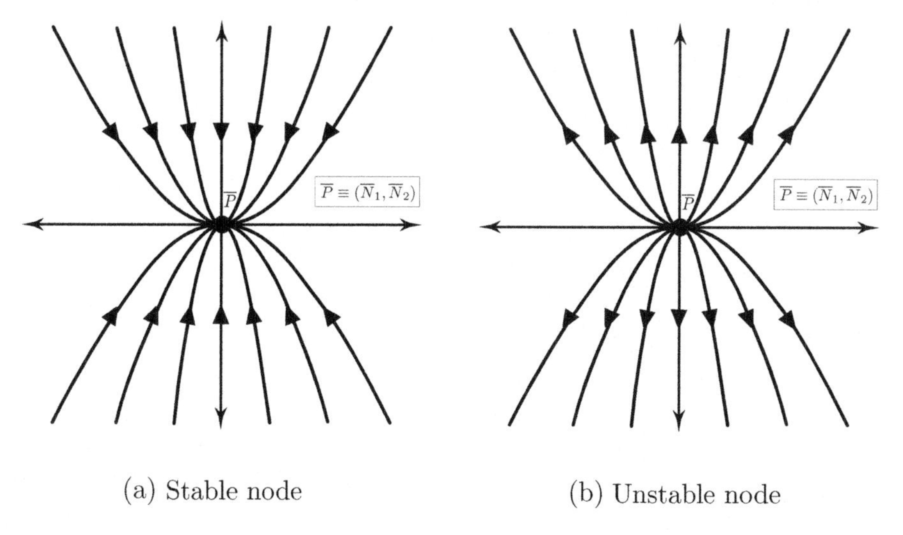

Fig. 1.4 Nodal points

where $a_i \in \mathbb{R}$ (for $i = 1, \ldots, 4$). Using $N_1(t) = \overline{N}_1 + u_1(t)$ and $N_2(t) = \overline{N}_2 + u_2(t)$ and (1.18), we have the following cases:

Case 1. NODE Let the roots λ_1 and λ_2 of the characteristic Eq. (1.17) are real and of the same sign. In this case, the equilibrium point $(\overline{N}_1, \overline{N}_2)$ is classified as a *node* (or *nodal point*). It has the following sub-cases:

• STABLE. If $\lambda_1, \lambda_2 < 0$, then the equilibrium point $(\overline{N}_1, \overline{N}_2)$ is a *stable node*. In this situation, $u_1(t) \to 0,\ u_2(t) \to 0$ as $t \to \infty$ and so $N_1(t) \to \overline{N}_1,\ N_2(t) \to \overline{N}_2$ as $t \to \infty$ (s Fig. 1.4a).

• UNSTABLE. If $\lambda_1, \lambda_2 > 0$, then the equilibrium point $(\overline{N}_1, \overline{N}_2)$ is an *unstable node*. In this situation, $u_1(t) \to \infty,\ u_2(t) \to \infty$ as $t \to \infty$ and so $N_1(t) \to \infty,\ N_2(t) \to \infty$ as $t \to \infty$ (see Fig. 1.4b).

Case 2. SADDLE POINT Let the roots λ_1 and λ_2 of the characteristic Eq. (1.17) are real and of the opposite sign. In this case, the equilibrium point $(\overline{N}_1, \overline{N}_2)$ is called a *saddle point*. Here $u_1(t) \to \infty,\ u_2(t) \to \infty$ as $t \to \infty$ and so $N_1(t) \to \infty,\ N_2(t) \to \infty$ as $t \to \infty$. Therefore, $(\overline{N}_1, \overline{N}_2)$ es *unstable* (see Fig. 1.5a).

Case 3. FOCUS Let the roots λ_1 and λ_2 of the characteristic Eq. (1.17) are complex conjugate numbers with nonzero real parts. In this case, the equilibrium point $(\overline{N}_1, \overline{N}_2)$ is known as *focus* (or *focal point*). Near a focal point, the solution trajectory behaves as spiral. It has the following sub-cases:

• STABLE If the real parts of $\lambda_1,\ \lambda_2$ are negative, then the focal point $(\overline{N}_1, \overline{N}_2)$ is *stable* and the solution trajectories spiral towards the equilibrium point $(\overline{N}_1, \overline{N}_2)$ as $t \to \infty$ (see Fig. 1.6a).

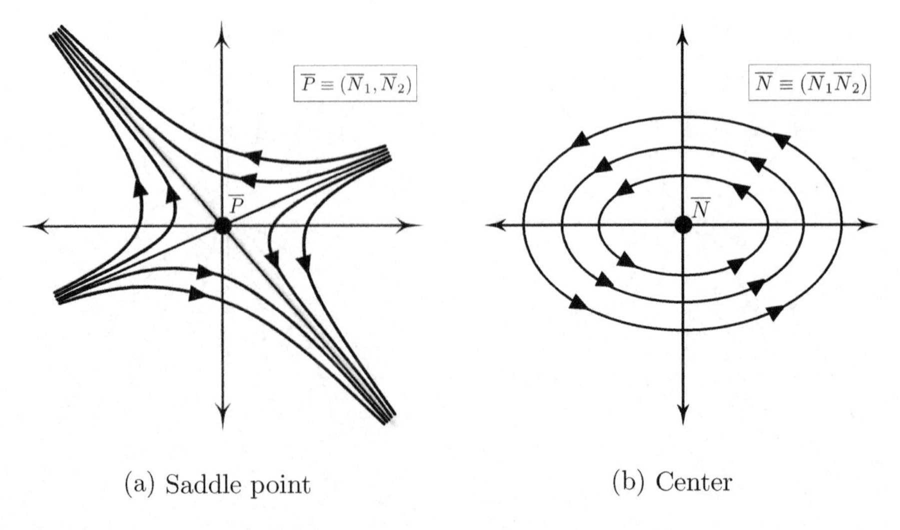

Fig. 1.5 **a** Saddle point, **b** Center

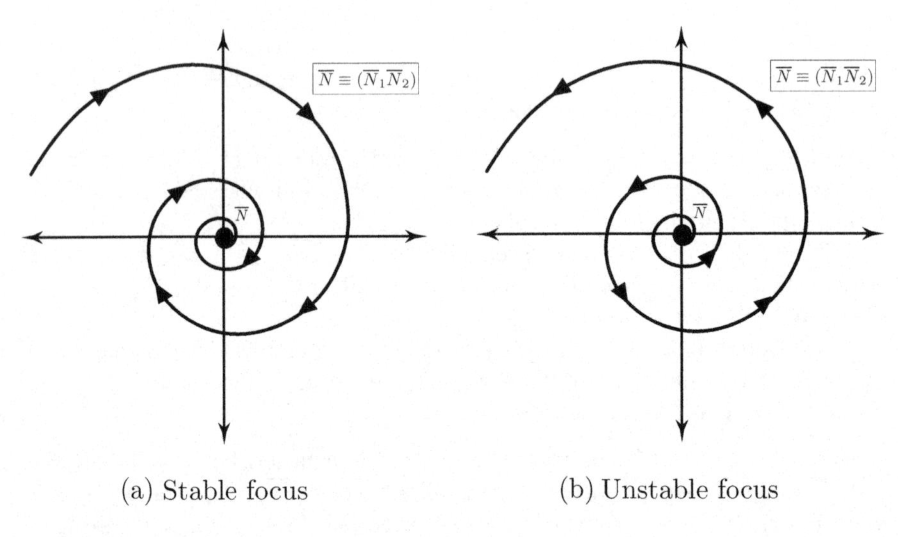

Fig. 1.6 Focal points

- UNSTABLE If the real parts of $\lambda_1,\ \lambda_2$ are positive, then the focal point $(\overline{N}_1, \overline{N}_2)$ is *unstable* and the solution trajectories spiral away from the equilibrium point as time goes (see Fig. 1.6b).

Case 4. CENTER Let the roots λ_1 and λ_2 of the characteristic Eq. (1.17) are pure imaginary numbers. In this case, the trajectories in the neighbourhood of the equilibrium point $(\overline{N}_1, \overline{N}_2)$ are closed curves and correspond to periodic solutions of the linear differential equations (1.15). Such an equilibrium point $(\overline{N}_1, \overline{N}_2)$ is known as *center* (or *vortex point*) and is illustrated in Fig. 1.5b. It is noted that centre is a stable equilibrium point, but not locally asymptotically stable.

1.4.3 *Linear (or Local) Stability of the Lotka–Volterra Model*

We know that the Lotka–Volterra model (1.6) has two equilibrium points, namely $\mathbf{0} \equiv (0, 0)$ y $P^* = \left(\frac{r_2}{b_2}, \frac{r_1}{b_1}\right)$. Let

$$f_1(N_1, N_2) = r_1 N_1 - b_1 N_1 N_2 \quad \text{and} \quad f_2(N_1, N_2) = -r_2 N_2 + b_2 N_1 N_2 \tag{1.19}$$

where $r_i, b_i > 0$ (for $i = 1, 2$) are positive constants . Therefore,

$$\begin{aligned} \frac{\partial f_1}{\partial N_1} &= r_1 - b_1 N_2 & \frac{\partial f_1}{\partial N_2} &= -b_1 N_1 \\ \frac{\partial f_2}{\partial N_1} &= b_2 N_2 & \frac{\partial f_2}{\partial N_2} &= -r_2 + b_2 N_1. \end{aligned} \tag{1.20}$$

I. Equilibrium point $(0, 0)$. In this case,

$$\left.\frac{\partial f_1}{\partial N_1}\right|_{(0,0)} = r_1, \quad \left.\frac{\partial f_1}{\partial N_2}\right|_{(0,0)} = 0, \quad \left.\frac{\partial f_2}{\partial N_1}\right|_{(0,0)} = 0, \quad \left.\frac{\partial f_2}{\partial N_2}\right|_{(0,0)} = -r_2.$$

Therefore, the secular (or characteristic) Eq. (1.17) becomes

$$\lambda^2 - (r_1 - r_2)\lambda - r_1 r_2 = 0 \implies (\lambda - r_1)(\lambda + r_2) = 0 \implies \lambda = r_1, -r_2.$$

The roots of the secular equation are real and of opposite signs and so the equilibrium point $(0, 0)$ is *unstable*, which is a *saddle point*.
II. Equilibrium point $P^* \equiv \left(\frac{r_2}{b_2}, \frac{r_1}{b_2}\right)$. Here

$$\left.\frac{\partial f_1}{\partial N_1}\right|_{P^*} = 0, \quad \left.\frac{\partial f_1}{\partial N_2}\right|_{P^*} = -\frac{b_1 r_2}{b_2}, \quad \left.\frac{\partial f_2}{\partial N_1}\right|_{P^*} = \frac{b_2 r_1}{b_1}, \quad \left.\frac{\partial f_2}{\partial N_2}\right|_{P^*} = 0.$$

Therefore, the secular equation is

$$\lambda^2 + r_1 r_2 = 0 \implies \lambda = \pm i\sqrt{r_1 r_2} \quad (\text{where } i = \sqrt{-1}).$$

Hence, the equilibrium point P^* is stable but not asymptotically stable, which is a center.
Conclusion: From the previous discussions, it is clear that the "interior", or the *coexistence* equilibrium point $P^* \equiv \left(\frac{r_2}{b_2}, \frac{r_1}{b_1}\right)$ of the Lotka–Volterra model (1.6) is not asymptotically stable. The absence of asymptotic stability of equilibrium point $P^* \equiv \left(\frac{r_2}{b_2}, \frac{r_1}{b_1}\right)$ indicates that *the Lotka-Volterra system has no mechanism to maintain its coexistence steady state. From an ecological point of view, the cause behind this is the absence of the concept of carrying capacity of the habitat for the prey species*

(i.e. the absence of intraspecific competition among the habitat of the prey species). From the view point of stability theory, the *coexistence* steady state $\left(\frac{r_2}{b_2}, \frac{r_1}{b_1}\right)$ *is a state of neutral equilibrium.*

1.5 Stabilization of Predator–Prey System by Introduction of Intraspecific Competition Among the Prey

The introduction intraspecific competition among the prey species, resulting from the limited resources, makes the Lotka–Volterra model in the following form:

$$\frac{dN_1}{dt} = r_1N_1 - \beta N_1N_2 - \gamma N_1^2$$
$$\frac{dN_2}{dt} = -r_2N_2 + \kappa\beta N_1N_2 \tag{1.21}$$

where the term γN_1^2 stands for intraspecifc competition among the prey species. It is immediately apparent that in the absence of the predators, the limit value of the prey population will be $\widetilde{N}_1 = \frac{r_1}{\gamma}$. The system (1.21) has a unique interior equilibrium at the point $N^* \equiv (N_1^*, N_2^*)$, where $N_1^* = \frac{r_2}{\kappa\beta}$ and $N_2^* = \frac{r_1\kappa\beta - \gamma r_2}{\kappa\beta^2}$. The realization of the natural condition $\widetilde{N}_1 > N_1^*$, that is $\frac{r_1}{\gamma} > \frac{r_2}{\kappa\beta}$, makes $N_2^* > 0$, and hence, the coexistence (or interior) equilibrium point (N_1^*, N_2^*) exists. The system (1.21) has other two equilibrium points, namely $(0, 0)$ and $\left(\frac{r_1}{\gamma}, 0\right)$.

The model (1.21) is also called *Logistic Lotka–Volterra predator–prey model*, or *Damped Lotka–Volterra predator–prey model.*

1.5.1 *Linear Stability Analysis of Logistic Lotka–Volterra Predator–Prey Model*

Let $g_1(N_1, N_2) = r_1N_1 - \beta N_1N_2 - \gamma N_1^2$ and $g_2(N_1, N_2) = -r_2N_2 + \kappa\beta N_1N_2$,

Therefore,

$$\frac{\partial g_1}{\partial N_1} = r_1 - \beta N_2 - 2\gamma N_1, \quad \frac{\partial g_1}{\partial N_2} = -\beta N_1, \quad \frac{\partial g_2}{\partial N_1} = \kappa\beta N_2, \quad \frac{\partial g_2}{\partial N_2} = -r_2 + \kappa\beta N_1.$$

I. Equilibrium point $(0, 0)$. In this case,

$$\left.\frac{\partial g_1}{\partial N_1}\right|_{(0,0)} = r_1, \quad \left.\frac{\partial g_1}{\partial N_2}\right|_{(0,0)} = 0, \quad \left.\frac{\partial g_2}{\partial N_1}\right|_{(0,0)} = 0, \quad \left.\frac{\partial g_2}{\partial N_2}\right|_{(0,0)} = -r_2,$$

Therefore, the secular Eq. (1.17) becomes

$$\lambda^2 - (r_1 - r_2)\lambda - r_1 r_2 = 0 \implies \lambda = r_1, -r_2.$$

Hence, the equilibrium point $(0, 0)$ is *unstable*, which is a *saddle point*.
II. Equilibrium point $\widetilde{N} \equiv \left(\frac{r_1}{\gamma}, 0\right)$. In this case,

$$dond \left.\frac{\partial g_1}{\partial N_1}\right|_{\widetilde{N}} = -r_1, \quad \left.\frac{\partial g_1}{\partial N_2}\right|_{\widetilde{N}} = -\frac{r_1\beta}{\gamma}, \quad \left.\frac{\partial g_2}{\partial N_1}\right|_{\widetilde{N}} = 0, \quad \left.\frac{\partial g_2}{\partial N_2}\right|_{\widetilde{N}} = \frac{\kappa\beta r_1 - r_2\gamma}{\gamma},$$

So, the secular Eq. (1.17) becomes

$$\lambda^2 - \left(-r_1 + \frac{\kappa\beta r_1 - r_2\gamma}{\gamma}\right)\lambda - \frac{r_1(\kappa\beta r_1 - r_2\gamma)}{\gamma} = 0$$

$$\Rightarrow \lambda = -r_1 \ (< 0), \quad \frac{\kappa\beta r_1 - r_2\gamma}{\gamma} \ (> 0).$$

Therefore, the equilibrium point $\left(\frac{r_1}{\gamma}, 0\right)$ is *unstable*, which is a *saddle point*.
III. Equilibrium point $N^* \equiv \left(N_1^*, N_2^*\right)$. In this case

$$dond \left.\frac{\partial g_1}{\partial N_1}\right|_{N^*} = -\gamma N_1^*, \quad \left.\frac{\partial g_1}{\partial N_2}\right|_{N^*} = -\beta N_1^*, \quad \left.\frac{\partial g_2}{\partial N_1}\right|_{N^*} = \kappa\beta N_2^*, \quad \left.\frac{\partial g_2}{\partial N_2}\right|_{N^*} = 0,$$

Therefore, the secular Eq. (1.17) becomes

$$\lambda^2 + \gamma N_1^*\lambda + \kappa\beta^2 N_1^* N_2^* = 0$$

$$\Rightarrow \lambda = \frac{1}{2}\left(-\gamma N_1^* \pm \sqrt{\Delta}\right), \quad \text{where} \quad \Delta = \gamma^2 (N_1^*)^2 - 4\kappa\beta^2 N_1^* N_2^*.$$

Therefore, either both roots of the secular equation are real and negative, or it has a pair of complex conjugate roots with negative real parts. Hence, the equilibrium point $\left(N_1^*, N_2^*\right)$ is *locally asymptotically stable*. It is known as the coexistence (or interior) equilibrium point.

1.5.2 Global Stability

Now, the question still remains: Is every trajectory initiating inside the positive quadrant eventually arrives at the coexistence equilibrium state $N^* \equiv (N_1^*, N_2^*)$? This implies *global asymptotic stability* of the state N^*, if the answer is 'yes'.

To search the answer of this question, let us first state a theorem:

Theorem 1.1 (LYAPUNOV'S THEOREM ON STABILITY) *Consider the following system of ordinary differential equations:*

$$\frac{dx_i}{dt} = f_i\left(x_1, x_2, \ldots, x_n\right), \quad i = 1, 2, \ldots, n \tag{1.22}$$

and assume that it has an equilibrium point at $(\overline{x}_1, \overline{x}_2, \ldots, \overline{x}_n)$. *If there exists a differentiable function* $v(x_1, x_2, \ldots, x_n)$ *that satisfies the following conditions:*

1. $v(x_1, x_2, \ldots, x_n)$ *has a strict minimum at* $(\overline{x}_1, \overline{x}_2, \ldots, \overline{x}_n)$, *that is* $v \geq 0$ *and* $v = 0$ *for* $x_i = \overline{x}_i$, $i = 1, 2, \ldots, n$.
2. *The derivative of* v *computed along the integral curve of (1.22) satisfies*

$$\frac{dv}{dt} = \sum_{i=1}^{n} \frac{\partial v}{\partial x_i}\frac{dx_i}{dt} = \sum_{i=1}^{n} \frac{\partial v}{\partial x_i} f_i \leq 0$$

and outside an arbitrarily small neighbourhood of $(\overline{x}_1, \overline{x}_2, \ldots, \overline{x}_n)$, $\dfrac{dv}{dt} < 0$,

then the equilibrium point $(\overline{x}_1, \overline{x}_2, \ldots, \overline{x}_n)$ *is* globally asymptotically stable. *The function* $v(x_1, x_2, \ldots, x_n)$ *so constructed is called* Lyapunov function. □

Let us state and prove an inequality which will be useful here.

Proposition 1.1 $z - \log(z) - 1 \geq 0$ *for all* $z > 0$, *and the equality occurs only for* $z = 1$.

Proof Set $w = z - \log(z) - 1$; therefore, $\frac{dw}{dz} = 1 - \frac{1}{z}$.

For $0 < z < 1$, we have $\frac{dw}{dz} < 0$, so w decreases as z increases. Hence, $w = z - \log z - 1 > 1 - \log 1 - 1 = 0$.

For $z > 1$, we have $\frac{dw}{dz} > 0$, so w increases as z increases. Hence, $w = z - \log(z) - 1 > 1 - \log(1) - 1 = 0$. □

On the basis of Lyapunov theorem and the Proposition 1.1, we set up the following function:

$$L(N_1, N_2) = N_1^* \left(\frac{N_1}{N_1^*} - \log \frac{N_1}{N_1^*} - 1 \right) + \frac{N_2^*}{\kappa} \left(\frac{N_2}{N_2^*} - \log \frac{N_2}{N_2^*} - 1 \right), \tag{1.23}$$

$\therefore L(N_1, N_2) \geq 0$ everywhere in the positive quadrant of the N_1N_2-plane and $L(N_1, N_2) = 0$ only at (N_1^*, N_2^*). Also,

$$\frac{dL}{dt} = \frac{\partial L}{\partial N_1}\frac{dN_1}{dt} + \frac{\partial L}{\partial N_2}\frac{dN_2}{dt}$$

$$= \left(1 - \frac{N_1^*}{N_1}\right)\left(r_1N_1 - \beta N_1N_2 - \gamma N_1^2\right)$$

$$+\frac{1}{\kappa}\left(1 - \frac{N_2^*}{N_2}\right)(-r_2N_2 + \kappa\beta N_1N_2)$$

$$\therefore \frac{dL}{dt} = -\gamma N_1^2 + \frac{N_2}{\kappa}\left(\kappa\beta N_1^* - r_2\right) + \left(r_1 + \gamma N_1^* - \beta N_2^*\right)N_1 - r_1N_1^* + \frac{r_2}{\kappa}N_2^*. \tag{1.24}$$

Now, $r_1 - \beta N_2^* - \gamma N_1^* = 0$, $-r_2 + \kappa\beta N_1^* = 0$, $-r_1N_1^* + \frac{r_2}{\kappa}N_2^* = -\gamma(N_1^*)^2$, and so

from (1.24) we get, $\frac{dL}{dt} = -\gamma\left(N_1 - N_1^*\right)^2 \leq 0$ and outside an arbitrarily small neighbourhood of (N_1^*, N_2^*), $\frac{dL}{dt} < 0$. Hence, $L(N_1, N_2)$ given by (1.23) is the corresponding Lyapunov function and the coexistence equilibrium state (N_1^*, N_2^*) is globally asymptotically stable.

1.6 Predator Functional Response on Prey Population

From the first equation of (1.21), it is observed that the number of prey eaten per predator per unit time is βN_1. Here, $\phi(N_1) = \beta N_1$ is called the *predator response function on prey population*, or *Volterra functional response*, or *Holling type I functional response*. So, the Holling type I response assumes that the predator is always hungry and that each individual of predator species eats more prey if more prey is available.

A more realistic predator response accounts for the predator species (say, foxes) eventually getting full. After all, how many individuals of prey species (say, rabbits) can one fox eat in unit time, really? As the prey (rabbit) population increases, the number of rabbits eaten by each fox increases at a decreasing rate. To get a reasonable mathematical answer, let us do the following analysis:

- Predation involves in two activities:
    1. Searching for prey (see Fig. 1.7a).
    2. Handling the prey (see Fig. 1.7b), where handling time refers to the time taken by a fox to pursue, capture and eat one rabbit.

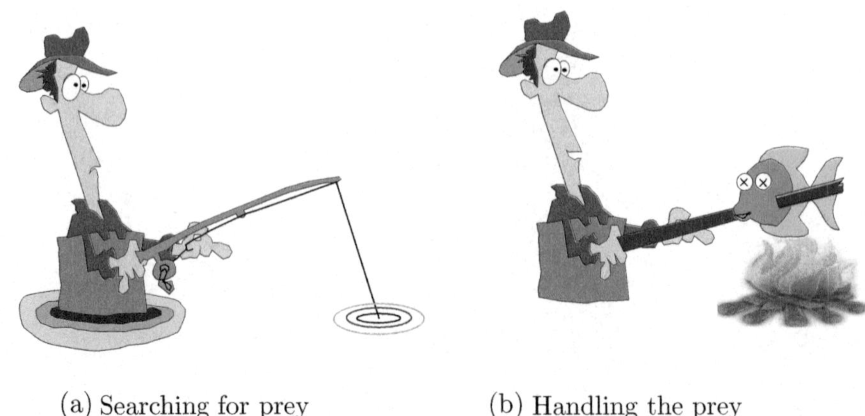

(a) Searching for prey (b) Handling the prey

Fig. 1.7 Searching for prey and handling the prey

- Total time available for predation is T.
- N_1 is the total number of prey (i.e. prey population size).
- V is the number of prey captured by one predator in unit time.
- T_h is the time taken to handle one prey per predator.
- $T - VT_h$ is the total searching time available for one predator.
- V is proportional to total prey population and total searching time $\Rightarrow V = \ell N_1(T - VT_h) \Rightarrow V = \dfrac{\ell T N_1}{1 + \ell T_h N_1}$, where ℓ is a constant.

So, a more realistic predator response function can be taken as

$$\psi(N_1) = \frac{\beta N_1}{\alpha + N_1}, \text{ where } \alpha, \beta > 0 \text{ are positive constants.}$$

This is called *Holling type II response* which is illustrated in Fig. 1.8.

Now, $\psi(N_1) \to \beta$ as $N_1 \to \infty$, and so the consumption cannot get arbitrarily large. It is observed that $\psi(\alpha) = \dfrac{\beta}{2}$. Here, α is called *half-saturation constant.*

Note: If total time available for predation is used only for searching (i.e. $T_h = 0$), then type I response is obtained.

If we incorporate Holling type II response in logistic Lotka–Volterra model, then we get

$$\begin{aligned} \frac{\mathrm{d}N_1}{\mathrm{d}t} &= rN_1\left(1 - \frac{N_1}{K}\right) - \frac{\beta N_1 N_2}{\alpha + N_1} \\ \frac{\mathrm{d}N_2}{\mathrm{d}t} &= -\mathrm{d}N_2 + \frac{c\beta N_1 N_2}{\alpha + N_1}. \end{aligned} \tag{1.25}$$

This known as *prey-dependent predator–prey system.*

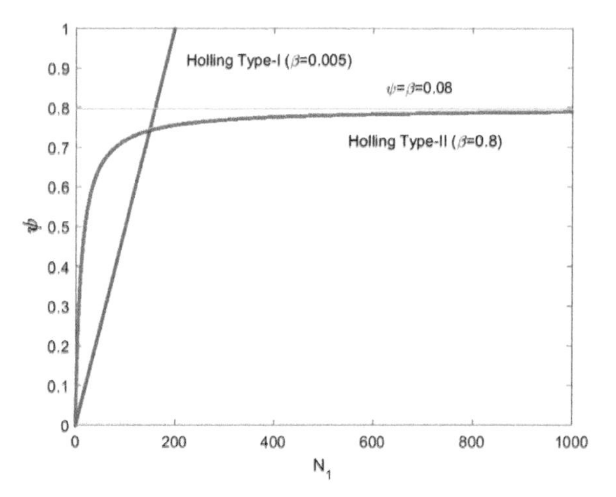

Fig. 1.8 Graphs of Holling type I and type II response

Instead of using $\psi(N_1)$, if we use $\psi\left(\frac{N_1}{N_2}\right) = \frac{\beta N_1}{\alpha N_2 + N_1}$ as predator response function on prey population in logistic Lotka–Volterra model, then we get

$$\begin{aligned}\frac{\mathrm{d}N_1}{\mathrm{d}t} &= rN_1\left(1 - \frac{N_1}{K}\right) - \frac{\beta N_1 N_2}{\alpha N_2 + N_1}\\ \frac{\mathrm{d}N_2}{\mathrm{d}t} &= -\mathrm{d}N_2 + \frac{c\beta N_1 N_2}{\alpha N_2 + N_1},\end{aligned} \tag{1.26}$$

where $\frac{\mathrm{d}N_1}{\mathrm{d}t} = \frac{\mathrm{d}N_2}{\mathrm{d}t} = 0$ for $(N_1, N_2) = (0, 0)$. This is known as *ratio-dependent predator–prey system.*

Notes: (i) The Holling type III response can be viewed as a generalization of the type II response to the form $\psi(N_1) = \frac{\beta N_1^k}{\alpha + N_1^k}$ which is not so easily formulated just by separating out the handling and searching behaviour. It has often been motivated by assuming that learning behaviour occurs in the predator population with a consequent increase in the discovery rate as more encounters with prey occur. Since at higher prey densities, there will be more previous encounters.

(ii) Holling types I, II and III functional responses are all monotonic in the first quadrant. It implies that, as the prey population increases, the consumption rate of prey per predator increases. However, there is experimental and observational evidence that indicates that this need not always be the case, for example in the case of **group defence** in population dynamics. **Group defence** is a term used to describe the phenomenon, whereby predation is decreased or even prevented altogether due to the increased ability of the prey to better defend or disguise themselves when their numbers are large enough. It is well known that large swarms of insects make individual identification difficult for their predators. To model such an effect, Andrews (1968) suggested a function $\psi(N_1) = \dfrac{\beta N_1}{\alpha + \gamma N_1 + N_1^2}$ called Holling type IV functional response.

1.7 Limit Cycles and Hopf Bifurcation

Consider the following system of differential equations:

$$\frac{dx}{dt} = y - \epsilon x(x^2 + y^2 - 1); \frac{dy}{dt} = -x - \epsilon y(x^2 + y^2 - 1) \tag{1.27}$$

where ϵ is a parameter.
Put $x = r\cos\theta, \quad y = rsin\theta$

$$\therefore \dot{x} = \dot{r}\cos\theta - r\dot{\theta}\sin\theta; \quad \dot{y} = \dot{r}\sin\theta + r\dot{\theta}\cos\theta$$

Therefore, (1.27) becomes:

$$\dot{r}\cos\theta - r\dot{\theta}\sin\theta = r\sin\theta - \epsilon r(r^2 - 1)\cos\theta \tag{1.28}$$

$$\dot{r}\sin\theta + r\dot{\theta}\cos\theta = -r\cos\theta - \epsilon r(r^2 - 1)\sin\theta \tag{1.29}$$

$$\left.\begin{array}{ll} (1.28)\times\cos\theta + (1.29)\times\sin\theta & \Rightarrow \dot{r} = -\epsilon r(r^2 - 1) \\ (1.28)\times(-\sin\theta) + (1.29)\times\cos\theta & \Rightarrow r\dot{\theta} = -r,\ or, \quad \dot{\theta} = -1 \end{array}\right\} \tag{1.30}$$

The functions $r(t) = 1$ and $\theta(\mathrm{t}) = -\mathrm{t}$ are solutions of (1.30), and so the circle $r = 1$ forms a closed orbit. Let $\epsilon > 0$. If $0 < r < 1$, then $\dot{r} > 0$, therefore $(0, 0)$ is unstable. For $r > 1$, $\dot{r} < 0$ and therefore, the closed orbit $r = 1$, i.e. $x^2 + y^2 = 1$ is stable.

Thus, every trajectory of (1.27) approaches the circle $x^2 + y^2 = 1$. The circle $x^2 + y^2 = 1$ is called a limit cycle of the system(1.1).

Definition 3 (Limit Cycle):

A closed trajectory of a dynamical system is called an orbit of the system. The motion along orbits is periodic. An orbit is said to be a limit cycle if every trajectory that starts at a point close to the orbit converges towards the orbit as $t \to \infty$.

According to the general theory of dynamical systems, any orbit that is not one of a family of concentric orbits must be either a limit cycle or an originating cycle in the sense that all neighbouring trajectories diverge from the orbit. An originating cycle is clearly a limit cycle for the time reversed dynamical system. Note that for $\epsilon < 0$, the unit circle $x^2 + y^2 = 1$ becomes an originating cycle for the system (1.27).

The secular equation of the system (1.27) is as follows:

$$\lambda^2 - \left(\frac{\partial f_1}{\partial x} + \frac{\partial f_2}{\partial y}\right)\lambda + \left(\frac{\partial f_1}{\partial x}\frac{\partial f_2}{\partial y} - \frac{\partial f_1}{\partial y}\frac{\partial f_2}{\partial x}\right) = 0$$

where

$$f_1(x, y) = y - \epsilon x(x^2 + y^2 - 1), \quad f_2(x, y) = -x - \epsilon y(x^2 + y^2 - 1)$$

$$\frac{\partial f_1}{\partial x} = -\epsilon(3x^2 + y^2 - 1); \quad \frac{\partial f_2}{\partial y} = -\epsilon(x^2 + 3y^2 - 1)$$

$$\frac{\partial f_1}{\partial y} = 1 - 2\epsilon xy; \quad \frac{\partial f_2}{\partial x} = -1 - 2\epsilon xy$$

Consider the equilibrium position $(0, 0)$. Here, the secular or characteristic equation is

$$\lambda^2 - 2\epsilon\lambda + \epsilon^2 + 1 = 0, \text{ or, } \quad (\lambda - \epsilon)^2 = -1$$

$$\therefore \lambda = \epsilon \pm i$$

Therefore, for $\epsilon = 0$, $(0, 0)$ is a centre.

If $\epsilon > 0$, $(0, 0)$ is unstable, and if $\epsilon < 0$, $(0, 0)$ is stable.

This is an example of **Hopf bifurcation** to occur at $\epsilon = 0$.

The **Hopf bifurcation** is characterized by a stability change of the equilibrium point (or critical point) accompanied by the creation of a limit cycle.

1.8 Jacobian Matrix or Variational Matrix

Given n functions $f_i(X_1, X_2, \ldots, X_n)$, $i = 1, 2, \ldots, n$ describing the dynamics of n variables X_i, $i = 1, 2, \ldots, n$, the *Jacobian matrix* or *variational matrix* J is defined as

$$J = \begin{pmatrix} \frac{\partial f_1}{\partial X_1} & \frac{\partial f_1}{\partial X_2} & \cdots & \frac{\partial f_1}{\partial X_n} \\ \frac{\partial f_2}{\partial X_1} & \frac{\partial f_2}{\partial X_2} & \cdots & \frac{\partial f_2}{\partial X_n} \\ \cdots & \cdots & \cdots & \cdots \\ \cdots & \cdots & \cdots & \cdots \\ \cdots & \cdots & \cdots & \cdots \\ \frac{\partial f_n}{\partial X_1} & \frac{\partial f_n}{\partial X_2} & \cdots & \frac{\partial f_n}{\partial X_n} \end{pmatrix}$$

To determine whether an equilibrium of interest is locally stable:
Step I: Evaluate the *Jacobian matrix* at the equilibrium of interest $J|_{\hat{X}_1,\hat{X}_2,\ldots,\hat{X}_n}$. This matrix is often called the local stability matrix.
Step II: Solve the characteristic polynomial $Det(J|_{\hat{X}_1,\hat{X}_2,\ldots,\hat{X}_n} - \lambda I) = 0$ (where I is an identity matrix of order n), which is an nth degree polynomial.
Step III: The n solutions to this characteristic polynomial are the n eigenvalues $\lambda_1, \lambda_2, \ldots, \lambda_n$.
Step IV: The real part of all n eigenvalues must be negative for the equilibrium to be locally stable. Equivalently, the equilibrium is locally stable if the real part of the eigenvalue with the largest real part (the leading eigenvalue) is negative.

1.9 Hopf Bifurcation in the Prey-Dependent Predator–Prey System

Let us rewrite the system (1.25) as:

$$\begin{aligned} \frac{dX}{dt} &= rX\left(1 - \frac{X}{K}\right) - \frac{\beta XY}{\alpha + X}, \quad X(0) > 0, \\ \frac{dY}{dt} &= -dY + \frac{c\beta XY}{\alpha + X}, \quad Y(0) > 0, \end{aligned} \tag{1.31}$$

where $X(t)$ and $Y(t)$ denote the prey and predator density, respectively, at time t; $r,\ K,\ d$ are positive constants that denote intrinsic growth rate of prey, carrying capacity of prey, and death rate of the predator, respectively, and $\alpha,\ \beta,\ c$ are positive constants that stand for the half capturing saturation constant, the capturing rate and the conversion rate, respectively.

1.9.1 Boundedness of the System

Boundedness of a system guarantees its validity. The following theorem ensures the boundedness of the system (1.31).

Theorem 1.2 *All solutions of the system (1.31) which start in $\mathbb{R}^2_+$ are uniformly bounded.*

Proof Let $(X(t), Y(t))$ be any solution of the system (1.31) with positive initial conditions. We have

$$X(t) = X(0)\exp\int_0^t \left[r\left\{1 - \frac{X(s)}{K}\right\} - \frac{\beta Y(s)}{\alpha + X(s)}\right] ds > 0,$$

$$Y(t) = Y(0)\exp\int_0^t \left\{-d + \frac{c\beta X(s)}{\alpha + X(s)}\right\} ds > 0.$$

Since $\frac{\mathrm{d}X}{\mathrm{d}t} \leq rX\left(1 - \frac{X}{K}\right)$, we have $\limsup_{t\to\infty} X(t) \leq \mu$. where $\mu = \max(X(0), K)$.

Let $W = cX + Y$ and so

$$\therefore \frac{\mathrm{d}W}{\mathrm{d}t} \leq crX\left(1 - \frac{X}{K}\right) - \mathrm{d}Y \leq 2cr\mu - \lambda W$$

$$\Rightarrow \frac{dW}{dt} + \lambda W \leq 2cr\mu, \text{ where } \lambda = \min(r, d).$$

Applying a theorem on differential inequalities (Birkhoff and Rota 1982), we obtain

$$0 < W(X, Y) \leq \left\{W(X(0), Y(0)) - \frac{2cr\mu}{\lambda}\right\} e^{-\lambda t} + \frac{2cr\mu}{\lambda}$$

and for $t \to \infty$,

$$0 < W \leq \frac{2cr\mu}{\lambda}.$$

Thus, all solutions of the system (1.31) enter into the region

$$B = \left\{(X, Y) : 0 < W < \frac{2cr\mu}{\lambda} + \epsilon, \text{ for any } \epsilon > 0\right\}$$

This proves the theorem. □

1.9.2 Equilibria

In all, the system (1.31) has three equilibrium points. $E_0(0, 0)$ is the trivial equilibrium, and $E_1(K, 0)$ is the only axial equilibrium. The interior equilibrium is $E^*(X^*, Y^*)$, where

$$X^* = \frac{\alpha d}{c\beta - d}, \quad Y^* = \frac{rc\alpha(Kc\beta - Kd - \alpha d)}{K(c\beta - d)^2}. \tag{1.32}$$

Existence of the interior equilibrium E^* depends upon the restriction

$$K(c\beta - d) > \alpha d.$$

The inequality (1.32) implies that the maximal benefit to the predator population from the interaction with the prey ($c\beta$) must exceed $\{1 + (\alpha/K)\}$ times the death rate of the predator (d) for the existence of the interior equilibrium point $E^*(X^*, Y^*)$, where K and α are the carrying capacity of the prey and the half-saturation constant, respectively.

1.9.3 Stability and Bifurcation Analysis

We now consider the stability analysis of the differential equations (1.31) governing the evolution of the system. The variational matrices corresponding to E_0 and E_1 are, respectively, given by

$$V(E_0) = \begin{bmatrix} r & 0 \\ 0 & -d \end{bmatrix},$$

$$V(E_1) = \begin{bmatrix} -r & -\frac{\beta K}{\alpha+K} \\ 0 & \frac{K(c\beta-d)-\alpha d}{\alpha+K} \end{bmatrix}.$$

Clearly, E_0 is unstable, and if $E^*(X^*, Y^*)$ exists, then E_1 is unstable. We now consider the stability of the most interesting positive equilibrium (interior equilibrium) $E^*(X^*, Y^*)$. The variational matrix at $E^*(X^*, Y^*)$ is given by

$$V(E^*) = \begin{bmatrix} r - \frac{2r}{K}X^* - \frac{\alpha\beta Y^*}{(\alpha+X^*)^2} & -\frac{\beta X^*}{\alpha+X^*} \\ \frac{c\alpha\beta Y^*}{(\alpha+X^*)^2} & 0 \end{bmatrix}. \tag{1.33}$$

Theorem 1.3 *If $E^*(X^*, Y^*)$ exists with $K(c\beta - d) < \alpha(c\beta + d)$, then E^* is locally asymptotically stable.*

Proof The characteristic equation for the variational matrix $V(E^*)$ (given by (1.33)) is

$$\lambda^2 - A_1\lambda + A_2 = 0,$$

where

$$A_1 = -\frac{r}{K}X^* + \frac{\beta X^* Y^*}{(\alpha + X^*)^2}, \quad \text{and } A_2 = \frac{c\alpha\beta^2 X^* Y^*}{(\alpha + X^*)^3} > 0.$$

Clearly, if $A_1 < 0$, then $E^*(X^*, Y^*)$ is locally asymptotically stable.

Now,

$$\begin{aligned} A_1 < 0 &\Longrightarrow -\frac{r}{K} + \frac{\beta Y^*}{(\alpha + X^*)^2} < 0 \\ &\Longrightarrow \frac{K\beta Y^*}{(\alpha + X^*)^2} < r \\ &\Longrightarrow \frac{r(Kc\beta - Kd - \alpha d)}{c\alpha\beta} < r \\ &\Longrightarrow K(c\beta - d) < \alpha(c\beta + d). \end{aligned}$$

This proves the theorem. □

Now, we prove that under the conditions stated above $E^*(X^*, Y^*)$ is a global attractor in the positive (X, Y)-plane.

Theorem 1.4 *If $E^*(X^*, Y^*)$ is locally asymptotically stable, then it is globally stable in the interior of the first quadrant.*

Proof If possible, let there be a periodic orbit $\Gamma = (X(t), Y(t)),\ 0 \le t \le T$ with the enclosed region Ω and consider the variational matrix V about the periodic orbit,

$$V = \begin{bmatrix} r - \frac{2r}{K}X - \frac{\alpha\beta Y}{(\alpha+X)^2} & -\frac{\beta X}{\alpha+X} \\ \frac{c\alpha\beta Y}{(\alpha+X)^2} & -d + \frac{c\beta X}{\alpha+X} \end{bmatrix}_{X=X(t),\ Y=Y(t)}.$$

We compute

$$\Delta = \int_0^T \left[r - \frac{2r}{K}X(t) - \frac{\alpha\beta Y(t)}{(\alpha + X(t))^2} - d + \frac{c\beta X(t)}{\alpha + X(t)} \right] \mathrm{d}t.$$

From the Eq. (1.31), it follows that

$$\int_0^T \left(-d + \frac{c\beta X(t)}{\alpha + X(t)} \right) dt = 0.$$

Therefore,

$$\Delta = \int_0^T \left[r - \frac{2r}{K}X(t) - \frac{r\alpha}{\alpha + X(t)} \left\{ 1 - \frac{X(t)}{K} \right\} \right] \mathrm{d}t. \text{ (by (1.31))}$$

Let

$$f(X) = r - \frac{2r}{K}X - \frac{r\alpha}{\alpha + X}\left(1 - \frac{X}{K}\right).$$

Then

$$f(X^*) = r - \frac{2r}{K}X^* - \frac{r\alpha}{\alpha + X^*}\left(1 - \frac{X^*}{K}\right)$$
$$= \frac{rd}{Kc\beta(c\beta - d)}\{K(c\beta - d) - \alpha(c\beta + d)\} < 0,$$

since $E^*(X^*, Y^*)$ is locally asymptotically stable.
Now,

$$\Delta = \int_0^T f(X(t))\mathrm{d}t$$
$$= \int_0^T \{f(X(t)) - f(X^*)\}\mathrm{d}t + \int_0^T f(X^*)\mathrm{d}t.$$

After some calculations, we obtain

$$\int_0^T \{f(X(t)) - f(X^*)\}\mathrm{d}t$$
$$= \oint_\Gamma \left[-\frac{r}{c\alpha\beta K}\{2(\alpha + X)(\alpha + X^*) - \alpha(\alpha + K)\}\right]\frac{1}{Y}\mathrm{d}Y$$
$$= \int_\Omega \left(-\frac{r}{c\alpha\beta K}\right)\frac{1}{Y}\frac{\mathrm{d}}{\mathrm{d}X}\{2(\alpha + X)(\alpha + X^*) - \alpha(\alpha + K)\}\,\mathrm{d}X\,\mathrm{d}Y$$
$$= \int_\Omega \left(-\frac{r}{c\alpha\beta K}\right)\frac{1}{Y}\{2(\alpha + Y^*)\}\mathrm{d}X\,\mathrm{d}Y < 0.$$

Therefore, $\Delta < 0$ and the periodic orbit Γ is orbitally asymptotically stable (Hale 1969, Cheng et al. 1981). Since every periodic orbit is orbitally stable, there is a unique limit cycle. From the Poincaré–Bendixson theorem, it is impossible that a unique stable limit cycle encloses a stable equilibrium. This is the desired contradiction. Hence, there is no limit cycle and $E^*(X^*, Y^*)$ is globally stable. □

Note: *Poincaré-Bendixson theorem* states that if a trajectory is confined to a closed, bounded region and there are no critical points in the region, then the trajectory must eventually approach a closed orbit. □

The condition for the asymptotic stability of the interior equilibrium point $E^*(X^*, Y^*)$ implies that

$$\frac{K + \alpha}{K}d < c\beta < \frac{K + \alpha}{K - \alpha}d.$$

This means that the maximal benefit to the predator population from the interaction with the prey ($c\beta$) should lie in the interval $(d(K + \alpha)/K, d(K + \alpha)/(K - \alpha))$ so as to have a guarantee of the asymptotic stability (local and global) of the interior equilibrium $E^*(X^*, Y^*)$, where K, α and d are the carrying capacity of the prey, the half-saturation constant and the predator death rate, respectively. Such a restriction suggests a possible control mechanism for the predator–prey system.

In the following theorem, we deduce the condition for the existence of Hopf bifurcating small amplitude periodic solutions near $E^*(X^*, Y^*)$.

Theorem 1.5 *If*

$$K = K^* = \frac{\alpha(c\beta + d)}{c\beta - d},$$

then the system (1.31) exhibits a Hopf bifurcation near $E^*(X^*, Y^*)$.and for $t \to \infty$,

$$0 < W \leq \frac{2cr\mu}{\lambda}.$$

Proof If the condition of the theorem is satisfied, then we have

$$(i)\ \left[\mathrm{tr}V(E^*)\right]_{K=K^*} = \frac{dr}{Kc\beta}\left[K - \frac{\alpha(c\beta + d)}{c\beta - d}\right] = 0,$$

$$(ii)\ \left[\det V(E^*)\right]_{K=K^*} = \frac{c\alpha\beta^2 X^* Y^*}{(\alpha + X^*)^3} > 0,$$

(iii) when E^*exists with $K = K^*$, then the characteristic equation is $\lambda^2 + [\det V(E^*)]_{K=K^*} = 0$ whose roots are purely imaginary,

$$(iv)\ \frac{\mathrm{d}}{\mathrm{d}K}\left[\mathrm{tr}V(E^*)\right]_{K=K^*} = -\frac{r(c\beta - d)}{c\alpha\beta} \neq 0.$$

Hence, all the conditions of the Hopf bifurcation theorem (Hassard 1981, Murray 1993) are satisfied and consequently there exist small amplitude periodic solutions near $E^*(X^*, Y^*)$. □

1.9.4 Numerical Simulation

In this section, we have validated our analytical results through computer simulation of some solutions of the system (1.31). Let us choose the values of the parameters as shown in Fig. 1.9. Then, the condition of Theorem 1.4 is fulfilled and $E^*(X^*, Y^*) = (0.4615, 0.1864)$ is locally asymptotically stable. The phase portrait is shown in Fig. 1.9a. In this situation, the prey and predator population approach to their equilibrium values x^* and y^*, respectively, as time progresses (see Fig. 1.9b).

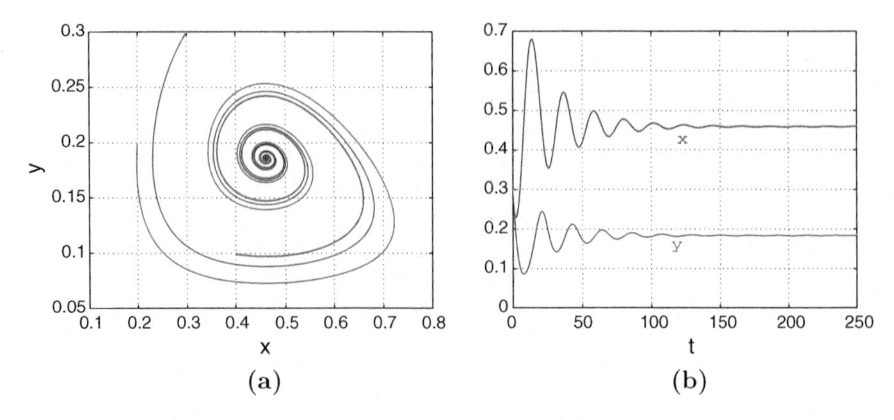

Fig. 1.9 Here, $r = 0.5,\ K = 1,\ \beta = 1.1,\ \alpha = 0.3,\ d = 0.8,\ c = 1.2$. **a** Phase portrait of the system (1.31) showing that $E^*(x^*, y^*) = (0.4615, 0.1864)$ is locally asymptotically stable. **b** The blue curve depicts the prey population, and the red curve depicts the predator population when $x(0) = 0.3, y(0) = 0.3$. Both the populations converge to their equilibrium-state values in finite time

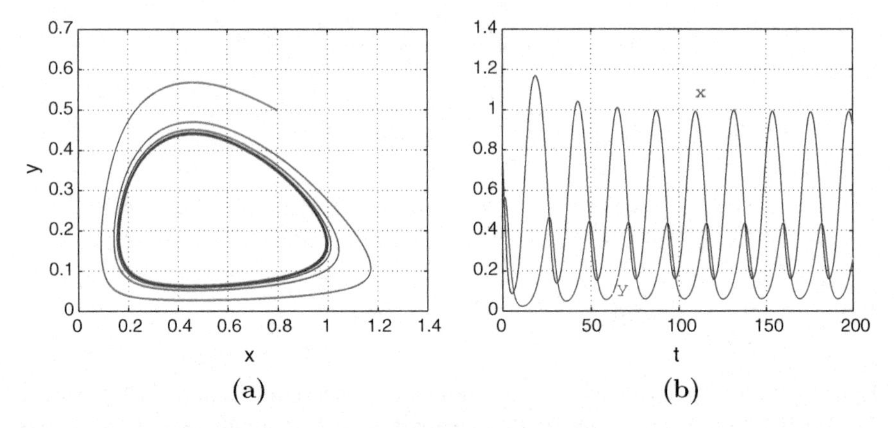

Fig. 1.10 Here, all the parameters are same as in Fig. 1.9 except $K = 1.4 > K^* = 1.22$. **a** Phase portrait of the system (1.31) showing a periodic orbit near $E^*(0.4615, 0.2320)$. **b** Oscillations of the prey and predator populations in time. (The blue curve represents the population density of the prey, and the red curve represents the population density of the predator.)

As the value of K increases, keeping the value of other parameters fixed, then by Theorem 1.5, there is a critical value $K = K^* = 1.22$ such that $E^*(X^*, Y^*)$ loses its stability as K passes through the critical value $K = K^* = 1.22$. For $K = 1.4 > K^*$, it is seen that $E^*(0.4615, 0.2320)$ is unstable and there is a periodic orbit near E^* (see Fig. 1.10a). The oscillations of prey and predator population are shown in Fig. 1.10b as time goes.

1.10 Hopf Bifurcation in the Ratio-Dependent Predator–Prey System

Let us rewrite system (1.26) as:

$$\begin{aligned} \frac{dX}{dd} &= X(a - bX) - \frac{cXY}{mY + X} = aX(1 - \frac{X}{K}) - \frac{cXY}{mY + X}, \\ \frac{dY}{dt} &= -dY + \frac{eXY}{mY + X}, \end{aligned} \tag{1.34}$$

where $X(t)$ and $Y(t)$, respectively, represent population biomass of prey and predator at time t, $K = a/b > 0$ is the carrying capacity of the prey, $d > 0$ is the death rate of the predator. Here, a, c, m and e are positive constants that stand for prey intrinsic growth rate, capturing rate, half-capturing saturation constant and conversion rate, respectively.

System (1.34) is undefined at (0, 0), and so we further assume that

$$X(0) > 0, \quad Y(0) > 0, \quad \text{and} \quad \frac{dX}{dt} = \frac{dY}{dt} = 0, \quad \text{for } (X, Y) = (0, 0).$$

We are interested in the dynamical behaviours of systems (1.34) in the interior of the first quadrant of the XY-plane where it is continuous and satisfies the Lipschitz condition.

1.10.1 Boundedness and Permanent of the System

The following theorem ensures the boundedness of the system (1.34).

Theorem 1.6 *All solutions of the system (1.34) which start in $\mathbb{R}^2_+$ are uniformly bounded.*

Proof Let $(X(t), Y(t))$ be any solution of the system (1.34) with positive initial conditions. We have

$$X(t) = X(0) \exp \int_0^t \left\{ a - bX(s) - \frac{cY(s)}{mY(s) + X(s)} \right\} ds > 0,$$

$$Y(t) = Y(0) \exp \int_0^t \left\{ -d + \frac{eX(s)}{mY(s) + X(s)} \right\} ds > 0.$$

Since $\frac{dX}{dt} \leq aX\left(1 - \frac{X}{K}\right)$, we have $\limsup_{t\to\infty} X(t) \leq \alpha$, where $\alpha = \max\{X(0), K\}$.

Let $B = eX + cY$ and so

$$\frac{dB}{dt} = aeX\left(1 - \frac{X}{K}\right) - cdY$$

$$\leq aeX - cdY = 2aeX - aeX - cdY$$

$$\leq 2ae\alpha - \beta B$$

$$\Rightarrow \frac{dB}{dt} + \beta B \leq 2ae\alpha, \text{ where } \beta = \min(a, d).$$

Applying a theorem on differential inequalities (Birkhoff and Rota 1982), we get

$$0 < B(X, Y) \leq \left\{B(X(0), Y(0)) - \frac{2ae\alpha}{\beta}\right\} e^{-\beta t} + \frac{2ae\alpha}{\beta}.$$

As $t \to \infty$, we have

$$0 < B \leq \frac{2ae\alpha}{\beta}.$$

Thus, all solutions of the system (1.34) enter into the region:

$$\Omega = \left\{(X, Y) : 0 < B < \frac{2ae\alpha}{\beta} + \epsilon, \text{ for any } \epsilon > 0\right\}.$$

This proves the theorem. □

Definition: A system represented by the following ordinary differential equations:

$$\frac{dx_i}{dt} = \phi_i(x_1, x_2, \ldots, x_n),\ i = 1, 2, \ldots, n$$

is said to be *permanent* (also called *uniformly persistent*) if there exists an $\eta > 0$ such that whenever $x_i(0) > 0$ for all i, $\liminf_{t\to\infty} x_i(t) > \eta$. In the ecological context, it implies the survival of all species which exist initially.

Theorem 1.7 *If $ma > c$ and $e > d$, then the system (1.34) is permanent.*

Proof We have

$$\frac{\mathrm{d}X}{\mathrm{d}t} = X(a - bX) - \frac{cXY}{mY + X}$$

$$= X(a - bX) - \frac{cX}{m}\frac{mY + X - X}{mY + X}$$

$$> X\left(a - \frac{c}{m} - bX\right).$$

$\therefore \liminf_{t\to\infty} X(t) \geq b^{-1}m^{-1}(ma - c)$ and so for large t, $X(t) > \widehat{X}/2$, where $\widehat{X} = b^{-1}m^{-1}(ma - c) > 0$, as $ma - c > 0$.

According to Theorem 1.6, there exists a positive constant M such that $0 < Y(t) \leq M$, $\forall t \geq 0$. Now for large t,

$$\frac{\mathrm{d}Y}{\mathrm{d}t} \geq Y\left\{-d + \frac{e\widehat{X}/2}{mY + (\widehat{X}/2)}\right\}$$

$$= Y\left\{(e - d)\frac{\widehat{X}}{2} - dmY\right\}\left(mY + \frac{\widehat{X}}{2}\right)^{-1}$$

$$> \left(mM + \frac{\widehat{X}}{2}\right)^{-1} Y\left\{(e - d)\frac{\widehat{X}}{2} - dmY\right\}$$

$$\Rightarrow \liminf_{t\to\infty} Y(t) \geq (e - d)\widehat{X}(2dm)^{-1} > 0, \text{ as } e > d.$$

Choosing $\eta > 0$ such that $\eta < \min\{\widehat{X}/2, \widehat{Y}/2\}$, where $\widehat{Y} = (e - d)\widehat{X}(2dm)^{-1} > 0$, we see that

$$\liminf_{t\to\infty} X(t) > \eta \text{ and } \liminf_{t\to\infty} Y(t) > \eta.$$

Hence, by definition, the system (1.34) is permanent. □

1.10.2 Equilibria

The system (1.34) always has the boundary equilibrium points $P_0(0, 0)$ and $P_1(a/b, 0)$. In this section, we are interested in the behaviour of the interior equilibrium point $P^*(X^*, Y^*)$. The interior equilibrium point $P^*(X^*, Y^*)$ exists uniquely if and only if any one of the following two conditions is true:

$$(i)\ d < e < \frac{cd}{c - ma}, \quad \text{when } c > ma,$$

$$(ii)\ e > d, \quad \text{when } c \leq ma.$$

In both the cases, X^* and Y^* are given by

$$X^* = \frac{e(am - c) + cd}{bme}, \quad Y^* = \frac{(e - d)\{e(am - c) + cd\}}{bdem^2}.$$

The system (1.34) cannot be linearized directly at $P_0(0, 0)$. A change $dt = (mY + X)d\tau$ in time scale transforms the model (1.34) into an equivalent polynomial system, and then, the stability of P_0 can be studied. Then, following Xiao and Ruan (2001), it can be observed that if $e \geq a + d$ and $c - am - dm \geq 0$, then P_0 is a global attractor for the system (1.34).

The variational matrix $V(P_1)$ at P_1 is given by

$$V(P_1) = \begin{bmatrix} -a & -c \\ 0 & e - d \end{bmatrix}.$$

Therefore, P_1 is locally asymptotically stable (or unstable) if $e < d$ (or $e > d$). Also, following Xiao and Ruan (2001), it is easy to see that if $e \leq d$ and $c - am - dm < 0$, then P_1 is a global attractor, and if $e \leq d$ with $c - am - dm \geq 0$, then P_1 is an attractor.

On the question of stability of $P^*(X^*, Y^*)$, we have the following theorem:

Theorem 1.8 *If P^* exists, then it is locally asymptotically stable or unstable according as* $\Delta = (c - am - dm)e^2 + (me - c)d^2 < or > 0.$

Proof The variational matrix at $P^*(X^*, Y^*)$ is as follows:

$$V(P^*) = \begin{bmatrix} -bX^* + \frac{cX^*Y^*}{(mY^*+X^*)^2} & -\frac{cX^{*2}}{(mY^*+X^*)^2} \\ \frac{emY^{*2}}{(mY^*+X^*)^2} & -\frac{emX^*Y^*}{(mY^*+X^*)^2} \end{bmatrix}.$$

The trace and determinant of $V(P^*)$ are as follows:

$$\begin{aligned} \mathrm{tr}V(P^*) &= -bX^* + (c - em)\frac{X^*Y^*}{(mY^* + X^*)^2} \\ &= \frac{(c - am - dm)e^2 + (me - c)d^2}{me^2}, \end{aligned}$$

$$\det V(P^*) = \frac{bemX^{*2}Y^*}{(mY^* + X^*)^2} > 0.$$

The characteristic equation for $V(P^*)$ is

$$\lambda^2 + E\lambda + Q = 0,$$

where $E = -\text{tr}V(P^*)$ and $Q = \det V(P^*)$.
It is clear that P^* is locally asymptotically stable or unstable according as $E >$ or < 0.
It completes the proof. □

A criterion for the existence of a Hopf bifurcation near $P^*(X^*, Y^*)$ is given in the following theorem:

Theorem 1.9 *If P^* exists with $c - m(a + d) > 0$, and*

$$e^* = \frac{-md^2 + \sqrt{m^2d^4 + 4cd^2(c - am - dm)}}{2(c - am - dm)},$$

then Hopf bifurcation occurs at $e = e^*$, provided $2c - me^* \neq 0$.

Proof It is noted that

(i) $\left[\text{tr}V(P^*)\right]_{e=e^*} = 0,$

(ii) $\left[\det V(P^*)\right]_{e=e^*} > 0,$

(iii) when P^*exists with $e = e^*$, the characteristic equation is
$\lambda^2 + [\det V(P^*)]_{e=e^*} = 0$ with purely imaginary roots,

(iv) $\dfrac{d}{de}\left[\text{tr}V(P^*)\right]_{e=e^*} = \dfrac{(2c - me^*)d^2}{me^3} \neq 0.$

Therefore, all the conditions of the Hopf bifurcation theorem (Murray 1993) are fulfilled and the theorem follows. □

Remark Occurrence of a Hopf bifurcation implies the existence of small amplitude periodic orbits near $P^*(X^*, Y^*)$.

1.10.3 Numerical Simulation

In this section, we perform computer simulation of some solutions of the system (1.34) to validate our analytical findings. These numerical simulation are very important from practical point of view (Figs. 1.11 and 1.12).

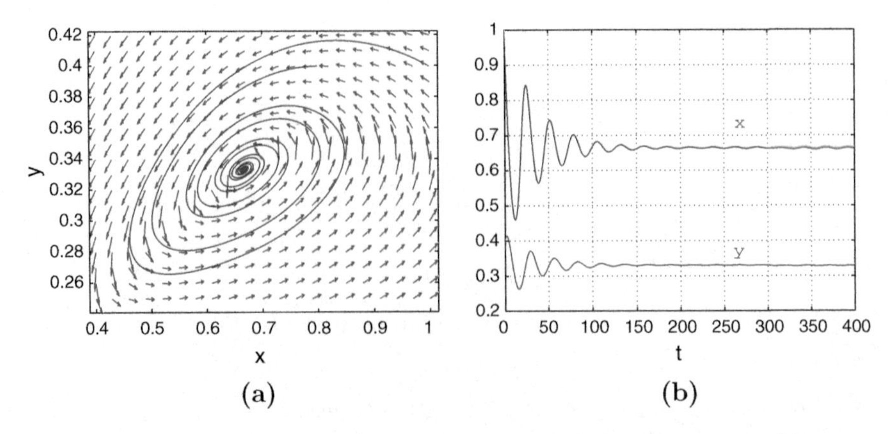

Fig. 1.11 Here $a = 1,\ b = 0.5,\ c = 2,\ m = 1,\ d = 0.5,\ e = 0.75$. **a** Phase portrait of the system (1.34). Clearly, it is a stable spiral converging to $P^*(X^*, Y^*) = (0.6667, 0.3333)$. **b** The blue curve depicts the prey population and the red curve one the predator population when $(X(0), Y(0)) = (1, 0.4)$. Both the population converge to their equilibrium-state values in finite time

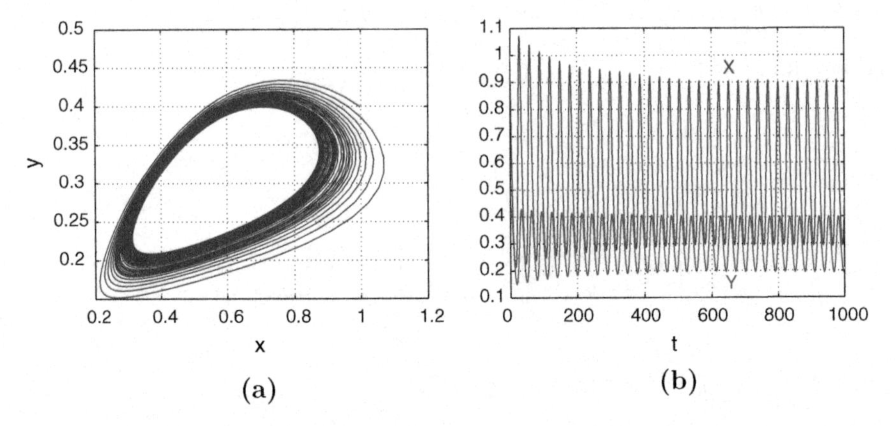

Fig. 1.12 **a** Phase portrait of the system (1.34) showing a periodic orbit near $P^*(X^*, Y^*) = (0.5575, 0.3145)$ when all the parameters are same as in Fig. 1.11 except $e = 0.782 > e^* = 0.7808$. **b** Oscillations of the prey and predator population in time when $(X(0), Y(0)) = (1, 0.4)$. The blue curve represents the population density of the prey, and the red curve represents the population density of the predator

Remark Thus, using the conversion rate e as control parameter, it is possible to break the stable behaviour of the system and drive it to an unstable state. Also, it is possible to keep the population at a desired level using the control e.

1.11 Leslie–Gower Predator–Prey Model

Leslie and Gower (1960) Leslie–Gower introduced the following predator–prey model, where the *carrying capacity* of the predator's environment is proportional to prey abundance:

$$\begin{aligned} \frac{dX}{dt} &= X(r_1 - a_1 Y - b_1 X), \quad X(0) > 0, \\ \frac{dY}{dt} &= Y\left(r_2 - a_2 \frac{Y}{X}\right), \quad Y(0) > 0, \end{aligned} \tag{1.35}$$

where $X(t)$ and $Y(t)$ denote the prey and predator density, respectively, at time t. All the constants in system (1.35) are positive. It is stressed the fact that there are upper limits to the rates of increase of both prey X and predator Y, which are not recognized in the Lotka–Volterra model. These upper limits can be approached under favourable conditions: for the predator, when the number of prey consumed per predator is large; for the prey, when the number of predator (and perhaps the number of prey also) is small.

1.11.1 Boundedness of the System

The following theorem ensures the boundedness of the system (1.35).

Theorem 1.10 *All solutions of the system (1.35) which start in $\mathbb{R}_+^2$ are bounded.*

Proof Let $(X(t), Y(t))$ be any solution of the system (1.35) with positive initial conditions. We have

$$X(t) = X(0) \exp \int_0^t \{r_1 - a_1 Y(s) - b_1 X(s)\}\, ds > 0,$$

$$Y(t) = Y(0) \exp \int_0^t \left\{ r_2 - a_2 \frac{Y(s)}{X(s)} \right\} ds > 0.$$

Since $\frac{dX}{dt} \le r_1 X \left(1 - \frac{X}{r_1/b_1}\right)$, we have $\limsup_{t\to\infty} X(t) \le \mu$. where $\mu = \max(X(0), r_1/b_1)$.

For large t:

$$\frac{dY}{dt} \le r_2 Y \left(1 - \frac{Y}{\mu r_2/a_2}\right) \Rightarrow \limsup_{t\to\infty} Y(t) \le \nu, \text{ where } \nu = \max\left\{Y(0), \frac{\mu r_2}{a_2}\right\}.$$

This proves the theorem. □

1.11.2 Stability Property of Interior Equilibrium

By simple computation, the system (1.35) admits an unique interior equilibrium point $E^*(X^*, Y^*)$, where

$$X^* = \frac{r_1 a_2}{r_2 a_1 + a_2 b_1}, \quad Y^* = \frac{r_1 r_2}{r_2 a_1 + a_2 b_1}.$$

Theorem 1.11 *The interior equilibrium point $E^*(X^*, Y^*)$ of system (1.35) is locally asymptotically stable.*

Proof The variational matrix at $E^*(X^*, Y^*)$ of system (1.35) is given by

$$V(E^*) = \begin{bmatrix} r_1 - a_1 Y^* - 2b_1 X^* & -a_1 X^* \\ a_2 \left(\frac{Y^*}{X^*}\right)^2 & r_2 - 2a_2 \frac{Y^*}{X^*} \end{bmatrix}. \tag{1.36}$$

Therefore, the characteristic equation for $V(E^*)$ is

$$\lambda^2 + P\lambda + Q = 0, \text{ where } P = r_2 + \frac{r_1 a_2 b_1}{r_2 a_1 + a_2 b_1}, \; Q = r_1 r_2 \tag{1.37}$$

It is clear that the roots of the characteristic equation are negative or have negative real parts. Hence, the unique interior equilibrium point $E^*(X^*, Y^*)$ of system (1.35) is locally asymptotically stable. □

Concerned with the global stability property of the interior equilibrium, we have the following:

Theorem 1.12 *The interior equilibrium point $E^*(X^*, Y^*)$ of system (1.35) is globally stable.*

Proof On the basis of Lyapunov theorem and the Proposition 1.1, we set up the following function:

$$L(X, Y) = \frac{X^*}{X} - \log\frac{X^*}{X} - 1 + \frac{a_1 X^*}{a_2}\left(\frac{Y^*}{Y} - \log\frac{Y^*}{Y} - 1\right), \tag{1.38}$$

$\therefore L(X, Y) \ge 0$ everywhere in the positive quadrant of the XY-plane and $L(X, Y) = 0$ only at (X^*, Y^*). Also,

$$
\begin{aligned}
\frac{dL}{dt} &= \frac{\partial L}{\partial X}\frac{dX}{dt} + \frac{\partial L}{\partial Y}\frac{dY}{dt} \\
&= \left(1 - \frac{X^*}{X}\right)(r_1 - b_1 X - a_1 Y) + \frac{a_1 X^*}{a_2}\left(1 - \frac{Y^*}{Y}\right)\left(r_2 - a_2\frac{Y}{X}\right) \\
&= \frac{X - X^*}{X}\left(b_1 X^* + a_1 Y^* - b_1 X - a_1 Y\right) + \frac{a_1 X^*}{a_2}\frac{(Y - Y^*)}{Y}\left(a_2\frac{Y^*}{X^*} - a_2\frac{Y}{X}\right) \\
&= -\frac{b_1}{X}(X - X^*)^2 - \frac{a_1}{Y}(Y - Y^*)^2 \le 0,
\end{aligned}
$$

and outside an arbitrarily small neighbourhood of (X^*, Y^*), $\frac{dL}{dt} < 0$. Hence, $L(X, Y)$ given by (1.38) is the corresponding Lyapunov function and the coexistence equilibrium state (X^*, Y^*) is globally asymptotically stable. □

The system (1.35) could not exhibit Hopf's bifurcation, and there is no limit cycle of this system in $\mathbb{R}_+^2$. To verify this, let us first state and prove the following theorem:

Theorem 1.13 (Bendixson–Du Lac Criterion): *Consider the following dynamical system:*

$$
\frac{dx}{dt} = F(x, y), \quad \frac{dy}{dt} = G(x, y) \tag{1.39}
$$

in which F(x,y) and G(x,y) are assumed to be smooth in a given simply connected region Ω. *Let B(x,y) be a smooth function in* Ω *such that the expression* $\frac{\partial}{\partial x}(BF) + \frac{\partial}{\partial y}(BG)$ *does not change sign in* Ω. *Then, the system (1.39) has no closed trajectory in* Ω.

Proof Suppose to the contrary that Γ is a closed trajectory in Ω. Then,

$$
I = \oint_{\Gamma}(-BG dx + BF dy) = \int_{\Omega_0}\left\{\frac{\partial}{\partial x}(BF) + \frac{\partial}{\partial y}(BG)\right\} dx dy \neq 0, \text{ by hypothesis,}
$$

where Ω_0 denotes the interior of Γ, and Green's theorem is employed in the second equality. But on the other hand, we have

$$
I = \int B\left(-G\frac{dx}{dt} + F\frac{dy}{dt}\right) dt = \int B(-GF + FG)dt = 0.
$$

This contradiction establishes the validity of the theorem. □

Let $F(X,Y) = X(r_1 - a_1 Y - b_1 X)$, $G(X,Y) = Y\left(r_2 - a_2\dfrac{Y}{X}\right)$ and $B(X,Y) = \dfrac{1}{XY}$.

Obviously, $F, G, B \in C^1(\mathbb{R}^2_+)$. Also, $\dfrac{\partial}{\partial X}(BF) + \dfrac{\partial}{\partial Y}(BG) = -\dfrac{b_1}{Y} - \dfrac{a_2}{X^2} < 0,\ \forall (X,Y) \in \mathbb{R}^2_+$.

According to Theorem 1.13, there is no limit cycle of the system (1.35) in $\mathbb{R}^2_+$.

1.12 Other Modifications

1.12.1 Time-Delay Effects

In the previous discussions, we have assumed that $\frac{dN_i}{dt}$, the rate of change of population size, depends on the instantaneous population size $N_i(t)$ only. However, we come across situations in which $\frac{dN_i}{dt}$ also depends on population sizes at earlier instants of time, i.e. on $N_i(t-\tau)$ with $\tau \geq 0$.

The response to growth rate may be delayed due to the following reasons::

1. Period of maturation (e.g. the period required for larvae to become adults).
2. Period of gestation (e.g. the period needed by predator to digest prey).

In the light of these, the model (1.26) can be modified as

$$\frac{dN_1}{dt} = rN_1\left(1 - \frac{N_1}{K}\right) - \frac{\beta N_1 N_2}{\alpha N_2 + N_1}$$

$$\frac{dN_2}{dt} = \left(-d + \frac{c\beta N_1(t-\tau)}{\alpha N_2(t-\tau) + N_1(t-\tau)}\right) N_2,$$

where $\frac{dN_1}{dt} = \frac{dN_2}{dt} = 0$ for $(N_1, N_2) = (0,0)$, with initial conditions $N_1(\theta) = \phi_1(\theta)$ and $N_2(\theta) = \phi_2(\theta)$ such that $\phi_i(\theta) \geq 0$ (for $i = 1,2$), for all $\theta \in [-\tau, 0]$, where $\phi_i(\theta)$ are continuous functions on $\theta \in [-\tau, 0]$. For a biological meaning, we further assume that $\phi_i(0) > 0$ (for $i = 1,2$).

1.12.2 Noise

Environmental fluctuations have a serious impact on real biological systems. Usually, such fluctuations do not strictly obey deterministic laws. It is observed that fluctuations will manifest themselves mainly as fluctuations in the intrinsic growth rate of the prey, r, and in the mortality rate of the predator, d, since these are the main parameters subject to coupling of a predator–prey pair with its environment.

Thus, the behaviour of the system (1.25) in a random environment will be considered within following framework:

$$\frac{dN_1}{dt} = \left(r + \eta_1(t) - \frac{r}{K}N_1\right)N_1 - \frac{\beta N_1 N_2}{\alpha + N_1}$$

$$\frac{dN_2}{dt} = (-d + \eta_2(t))\,N_2 + \frac{c\beta N_1 N_2}{\alpha + N_1},$$

where the perturbation terms $\eta_1(t)$ and $\eta_2(t)$ are assumed as the independent coloured noises (or Ornstein–Uhlenbeck processes), which are realistic noises as per experimental evidence. The mathematical expectations and the correlation functions of the stochastic processes $\eta_j(t)$ are as follows: $\langle \eta_j(t) \rangle = 0$, $\langle \eta_j(t_1)\eta_j(t_2) \rangle = \epsilon_j \delta_j \exp(-\delta_j |t_1 - t_2|)$ (for $j = 1, 2$), where ϵ_j (> 0) and δ_j^{-1} (> 0) are, respectively, the intensity and the correlation time of the noises $\eta_j(t)$ and $\langle \cdot \rangle$ represents the average over the ensemble of the stochastic process.

If $\delta_j^{-1} \to 0$, then $\eta_j(t) \to \sqrt{2\epsilon_j}\xi_j(t)$, where $\xi_j(t)$ are independent zero mean standard Gaussian white noises characterized by the properties: $\langle \xi_j(t) \rangle = 0$ and $\langle \xi_j(t_1)\xi_j(t_2) \rangle = \delta(t_1 - t_2)$ (for $j = 1, 2$), where $\delta(t)$ is the Dirac delta function defined by

$$\begin{cases} \delta(t) = 0, \ \text{for } t \neq 0, \\ \int\limits_{-\infty}^{\infty} \delta(t)dt = 1. \end{cases}$$

It is well known that Gaussian white noise is a very irregular random process with no memory at all (the correlation function is a Dirac delta function), which is extremely useful to model rapidly fluctuating phenomena. It is experimentally recognized that true white noises do not occur in nature. However, as can be seen by studying their spectra: thermal noise in electrical resistance, the force acting on a Brownian particle, climate fluctuations, disregarding the periodicity of astronomical origin, etc., are white to a very good approximation, which support the usefulness of the white noise idealization in applications to natural phenomena. Also, it is known that the process $(N_1(t), N_2(t))$ is *Markovian* if and only if the external noises are white. These observations explain the importance and appeal of the white noise idealization. It is also experimentally recognized that the coloured noises (or Ornstein–Uhlenbeck processes) are more realistic noises than white noises.

1.13 Kinetics of Growth and Ageing

The first developer of the thermodynamic theory of biological growth by using the phenomenological principles of non-equilibrium thermodynamics was Prigogine and Wiame (1946). However, as growth and ageing are internally stochastic processes irrespective of the random effect of the environment, both the principles of non equilibrium thermodynamics and the technique of stochastic process to include the effects of noise or fluctuation should be embraced by a proper theory of biological growth and ageing. Our aim in this section is to study the stochastic behaviour of biological growth and ageing using the stochastic analysis of non-equilibrium fluctuation of the growth and ageing parameter of the living system. The consideration of non-equilibrium thermodynamics gives the stochastic extension of the phenomenological kinetic equation which is the base of the theory.

1.13.1 Dissipative Function: Basic Kinetic Equation

A generalized thermodynamic theory of biological ageing of living system has been developed by Zotina and Zotin (1967). The subdivision of the dissipative function (the rate of change of entropy production) as the sum of two functions is the key of this theory:

$$\psi = \psi_u + \psi_d \tag{1.40}$$

where the bound dissipative function ψ_u tends to zero as the system tends to the local stationary state and the external dissipative function ψ_d satisfies the total criteria of evolution:

$$\frac{\mathrm{d}\psi_\mathrm{d}}{\mathrm{d}t} \leq 0 \tag{1.41}$$

which specifies the ageing of the system. Existence of a dissipative function ensures the damping of all fluctuations. That is the reason why near local stationary state, a macroscopic description for large systems is sufficient: fluctuations can only play a subordinate role, appearing as corrections to the macroscopic laws which can be neglected. On the basis of the subdivision (Eq. (1.40)), Timonin and Zotina (1983) have derived a general kinetic equation:

$$\frac{\mathrm{d}x}{\mathrm{d}t} = \phi_1(x)\psi_u(t) + \phi_2(x)\psi_\mathrm{d}(t) \tag{1.42}$$

which describes the law of evolution of the growth parameter x (such as weight and mass) of living and artificial systems. Equation (1.42) is the basic deterministic kinetic equation. Different types of growth such as tumour growth, embryonic growth, growth of animals and men are described by different forms of functions $\phi_1(x)$, $\phi_2(x)$, $\psi_u(t)$ and $\psi_d(t)$.

1.13.2 Kinetic of Biological Growth

Now, we consider an example where

$$\begin{aligned} \phi_1(x) = x, \phi_2(x) = 0 \\ \psi_u(t) = ae^{-\alpha t}, \psi_d(t) = 0 \end{aligned} \tag{1.43}$$

From the kinetic Eq. (1.42), we get

$$\frac{\mathrm{d}x}{\mathrm{d}t} = ae^{-\alpha t}x \tag{1.44}$$

which is the Gompertzian equation of growth. Tumour growth, embryonic growth, growth of animals and men are described by this type of equation.

However, the process of ageing of living system cannot be explained using Eq. (1.44). The kinetic behaviour of such cases can be studied by extending the Gompertzing equation (1.44) to the form:

$$\frac{\mathrm{d}x}{\mathrm{d}t} = ae^{-\alpha t}x + (be^{-\beta t} + c). \tag{1.45}$$

The expression $ae^{-\alpha t}$ in the above equation is the bound dissipative function. According to Timonin and Zotina (1983), this term tends to zero when the system evolves to the local stationary state. Thus, the rate of the system's drift to a local stationary point is described by the first term. The external dissipative function $(be^{-\beta t} + c)$ in the above equation describes the ageing effect of the living system. Equation (1.40) is an in-homogeneous non-autonomous equation governing the growth and ageing of living system.

The stochastic extension of the deterministic Eq. (1.45) is given by

$$\frac{\mathrm{d}x}{\mathrm{d}t} = ae^{-\alpha t}x + (be^{-\beta t} + c) + \sqrt{2\epsilon}\eta(t) \tag{1.46}$$

where $\eta(t) = \dfrac{\mathrm{d}W}{\mathrm{d}t}$ denotes the standard zero mean Gaussian white noise characterized by

$$\langle \eta(t) \rangle = 0, \quad \text{and} \quad \langle \eta(t_1)\eta(t_2) \rangle = \delta(t_1 - t_2), \tag{1.47}$$

$\langle \cdot \rangle$ represents the average over the ensemble of the stochastic process satisfying Eq. (1.46) with the diffusion coefficient ϵ. Here $W(t)$ is a unit Wiener process.

Equation (1.46) is a generalization of Langevin equation with a time dependent drift $A(t) = ae^{-\alpha t}$ and a time-dependent driving force $(be^{-\beta t} + c)$ describing the ageing effect of the living system. The equation of motion of the average $\langle x(t) \rangle$ is then given by

$$\frac{\mathrm{d}\langle x \rangle}{\mathrm{d}t} = ae^{-\alpha t}\langle x \rangle + (be^{-\beta t} + c). \tag{1.48}$$

Solving the equation, we get

$$\langle x(t)\rangle = \langle x_0\rangle \exp\left\{-\frac{a}{\alpha}(e^{-at}-1)\right\} + \int_0^t (be^{-\beta t}+c)\exp\left\{-\frac{a}{\alpha}(e^{-at}-1)\right\}dt$$

$$= \langle x_0\rangle \exp\left\{-\frac{a}{\alpha}(e^{-at}-1)\right\} + K(t),\ \text{where } K(t) = \int_0^t (be^{-\beta t}+c)\exp\left\{-\frac{a}{\alpha}(e^{-at}-1)\right\}dt \tag{1.49}$$

Now, we take $F(x,t) = x^2$, and applying the following *Itô* formula:

$$\mathrm{d}F(x,t) = \left(\frac{\mathrm{d}F}{\mathrm{d}t} + \frac{\mathrm{d}F}{\mathrm{d}x}f(x) + \frac{1}{2}\frac{\partial^2 F}{\partial x^2}g_{11}^2\right)\mathrm{d}t + \frac{\mathrm{d}F}{\mathrm{d}x}g_{11}dW$$

$$\therefore \mathrm{d}F = \left[2x(ae^{-\alpha t}x + be^{-\beta t}+c) + 2\epsilon\right]\mathrm{d}t + 2\sqrt{2\epsilon}x\,dW$$

$$\therefore \langle x^2\rangle = x_0^2 + 2a\langle x^2\rangle\int_0^t e^{-\alpha t}\mathrm{d}t + 2b\langle x\rangle\int_0^t e^{-\beta t}dt + 2c\langle x\rangle\int_0^t dt + 2\epsilon t$$

$$\Rightarrow \langle x^2\rangle\left(1-\frac{2a}{\alpha}+\frac{2a}{\alpha}e^{-\alpha t}\right) = x_0^2 + \frac{2b}{\beta}(1-e^{-\alpha t})\left[x_0e^{-\frac{a}{\alpha}(1-e^{-\alpha t})}+K(t)\right]$$

$$+2ct\left[x_0e^{-\frac{a}{\alpha}(1-e^{-\alpha t})}+K(t)\right]+2\epsilon t$$

$$\Rightarrow \langle x^2\rangle = \frac{x_0^2 + \frac{2b}{\beta}(1-e^{-\alpha t})\left[x_0e^{-\frac{a}{\alpha}(1-e^{-\alpha t})}+K(t)\right] + 2ct\left[x_0e^{-\frac{a}{\alpha}(1-e^{-\alpha t})}+K(t)\right]+2\epsilon t}{\left(1-\frac{2a}{\alpha}+\frac{2a}{\alpha}e^{-\alpha t}\right)}$$

$\therefore$ Variance of $x = \sigma^2(t) = \langle x^2\rangle - \langle x\rangle^2 =$

$$\frac{x_0^2 + \frac{2b}{\beta}(1-e^{-\alpha t})\left[x_0e^{-\frac{a}{\alpha}(1-e^{-\alpha t})}+K(t)\right] + 2ct\left[x_0e^{-\frac{a}{\alpha}(1-e^{-\alpha t})}+K(t)\right]+2\epsilon t}{\left(1-\frac{2a}{\alpha}+\frac{2a}{\alpha}e^{-\alpha t}\right)}$$

$$-x_0e^{-\frac{2a}{\alpha}(1-e^{-\alpha t})} - K(t)^2 - 2K(t)x_0e^{-\frac{a}{\alpha}(1-e^{-\alpha t})} \tag{1.50}$$

As $t\to\infty$,

$$\sigma^2(t) \to \frac{x_0^2 + \frac{2b}{\beta}(x_0 e^{\frac{a}{\alpha}} + L) + 2\left(\epsilon + c x_0 e^{\frac{a}{\alpha}} + cL\right) t}{1 - \frac{2a}{\alpha}} - x_0^2 e^{\frac{2a}{\alpha}} - L^2 - 2L x_0 e^{\frac{a}{\alpha}}$$

$$= 2\left(\frac{\epsilon + c x_0 e^{\frac{a}{\alpha}} + cL}{1 - \frac{2a}{\alpha}}\right) t + \left(\frac{x_0^2 + \frac{2b}{\beta}(x_0 e^{\frac{a}{\alpha}} + L)}{1 - \frac{2a}{\alpha}}\right) - x_0^2 e^{\frac{2a}{\alpha}} - L^2 - 2L x_0 e^{\frac{a}{\alpha}}$$

$$= Mt + N, \text{ where } L := \lim_{t\to\infty} \int_0^t (be^{-\beta t} + c) \exp\left\{-\frac{a}{\alpha}(e^{-at} - 1)\right\} dt \tag{1.51}$$

Therefore, in the limit of large time t, the non-stationary state near the steady state, the fluctuation of the growth parameter $x(t)$ varies linearly with time t. It leads to an interesting result on the behaviour of non-equilibrium fluctuation of the growth depending on the ageing parameters.

References

Andrews, J.F.: A mathematical model for the continuous culture of microorganisms utilizing inhibitory substrates. Biotechnol. Bioeng. **10**, 707–723 (1968)

Birkhoff, G., Rota, G.C.: Ordinary Differential Equations. Ginn, Boston (1982)

Cheng, K.S., Hsu, S.B., Lin, S.S.: Some results on global stability of a predator-prey system. J. Math. Biol. **12**, 115–126 (1981)

Hale, J.K.: Ordinary Differential Equations. Wiley-Interscience, New York (1969)

Hassard, B.D., Kazarinoff, N.D., Wan, Y.H.: Theory and Application of HopfHopf-bifurcation. Cambridge University Press, Cambridge (1981)

Leslie, P.H., Gower, J.C.: The properties of a stochastic model for the predator-prey type of interaction between two species. Biometrica **47**, 219–34 (1960)

Murray, J.D.: Mathematical Biology. Springer-Verlag, New York (1993)

Prigogine, I., Wiame, J.M.: Biologie et thermodynamique des phnomnes irrversibles. Experientia **2**(11), 451–453 (1946)

Timonin, V.I., Zotina, R.S.: Thermodynamics and Kinetics of Biological Processes. Walter-de-Gruyter, Berlin (1983)

Xiao, D., Ruan, S.: Global dynamics of a ratio-dependent predator-prey system. J. Mathe. Biol. **43**, 268–290 (2001)

Chapter 2
Dynamical Models of Single-Species System in a Polluted Environment

2.1 Introduction

From the viewpoints of mathematics and experiments: the effect of toxicants on ecosystem (a community of living organisms with the environment where they lives), due to change in both terrestrial and aquatic environments caused by various kinds of stresses, is an absolutely essential issue. It is of great interest from both environmental and conservational points of view. Degradation of forests (due to acid rain), dumping of toxic (hazardous) waste in rivers and lakes and the environmental fluctuation because of eutrophication in the Salton sea leading to pelican extinction are some examples. In this chapter, we have studied dynamical behaviours of two single-species population models in polluted environment.

2.2 Dynamical Model I of a Single-Species System in a Polluted Environment

In this article, we have developed a single-species population model in polluted environment, where the species carrying capacity is also affected by the exogenous introduction of toxicant. Two situations are analysed: (i) constant exogenous input of toxicant into the environment, and (ii) rapidly fluctuating random exogenous input of toxicant into the environment. The following mathematical techniques are used: (i) deterministic linearization, (ii) Lyapunov method and (iii) stochastic linearization (on the assumption that exogenous input of toxicant into the environment behaves like *Coloured noise*).

© Springer Nature Singapore Pte Ltd. 2021

G. Samanta, *Deterministic, Stochastic and Thermodynamic Modelling of some Interacting Species*, Forum for Interdisciplinary Mathematics,
https://doi.org/10.1007/978-981-16-6312-3_2

2.2.1 The Basic Mathematical Model

Consider the following population-toxicant model using a modified logistic equation (Freedman 1987):

$$\begin{aligned}\frac{\mathrm{d}X}{\mathrm{d}t} &= g(L)X - \frac{g_0X^2}{K(E)},\\ \frac{\mathrm{d}E}{\mathrm{d}t} &= -\zeta_0 E + e(t),\\ \frac{\mathrm{d}L}{\mathrm{d}t} &= -\zeta_1 L + a_1 E + \frac{d_1\alpha\beta}{a_1},\end{aligned} \tag{2.2.1}$$

where $X(t)$ represents the population density at time t, the organismal and environmental toxicant concentrations at time t are expressed by $L(t)$, $E(t)$, respectively, the exogenous input of toxicant into the environment is denoted by $e(t)$.
The following meanings of the parameters are adopted:
a_1: the population's rate of toxicant uptake from the environment per unit mass.
d_1: the uptake rate of toxicant in food per unit mass.
α: the concentration of toxicant in the resources.
β: the average rate of food intake per unit organismal mass.
ζ_1: the sum of net ingestion and depuration rates of organismal toxicant.
ζ_0: the loss rate of toxicant from the environment.
Assumptions:
1. $\zeta_0, \zeta_1, g_0, a_1, d_1, \alpha, \beta$ are all positive constants.
2. Growth of X is diminished by the presence of organismal toxicant L in X.
3. The carrying capacity of the environment is decreased because of the presence of environmental toxicant E.
4. $g(L)X$ is the birth component of population, and hence, $g(L)$ represents the growth rate affected by organismal toxicant L.
5. The death component is given by $g_0X^2/K(E)$, where $K(E)$ represents the carrying capacity which is affected by environmental toxicant E.
6.

$$g(0) = g_0 > 0,\ g'(L) < 0,\ \text{for } L \geq 0, \tag{2.2.2}$$

7.

$$K(0) = K_0 > 0,\ K'(E) < 0,\ \text{for } E \geq 0, \tag{2.2.3}$$

8.

$$\zeta_0 > a_1. \tag{2.2.4}$$

Conditions (2.2.2) and (2.2.3) indicate that increasing toxicant decreases birth rate and increases death rate, respectively, of the population.

9. $e(t)$ represents the rate of exogenous input of toxicant into the environment beyond the initial concentration. Our purpose is to analyse model (2.2.1) for three possible situations: (i) $e(t) = 0$, (ii) $e(t) = Q > 0$ and (iii) $e(t)$ is a rapidly fluctuating random perturbation.

2.2.2 *Dynamical Behaviour*

Situation I. Zero exogenous input: $e(t) = 0$.
We first study boundedness of the system (2.2.1) in this situation.

Theorem 2.1 *If $e(t) = 0$, then all solutions of system (2.2.1) starting in $\mathbb{R}_+^3$ are uniformly bounded.*

Proof It is observed that

$$\frac{\mathrm{d}X}{\mathrm{d}t} = g(L)X - \frac{g_0 X^2}{K(E)} \leq g_0\left(X - \frac{X^2}{K_0}\right).$$

Therefore, using a standard comparison theorem:

$$\limsup_{t\to\infty} X(t) \leq K_0.$$

Define:

$$G = E + L.$$

The time derivative of G along a solution of (2.2.1) is

$$\begin{aligned}\frac{\mathrm{d}G}{\mathrm{d}t} &= -(\zeta_0 - a_1)E - \zeta_1 L + b \\ &\leq -\zeta G + b, \quad \text{where } \zeta = \min\{(\zeta_0 - a_1), \zeta_1\}.\end{aligned}$$

$$\therefore \frac{\mathrm{d}G}{\mathrm{d}t} + \zeta G \leq b.$$

Applying a theorem on differential inequalities (Birkhoff and Rota 1982):

$$0 \leq G(E, L) \leq \frac{b}{\zeta} + \frac{G(E(0), L(0))}{\mathrm{e}^{\zeta t}}.$$

For $t \to \infty : 0 \leq G \leq \dfrac{b}{\zeta}$.

$$\therefore \overline{\Omega} = \left\{(X, E, L) : 0 \leq X \leq K_0,\ 0 \leq E + L \leq \frac{b}{\zeta},\ b = \frac{d_1 \alpha\beta}{a_1},\ \zeta = \min(\zeta_0 - a_1, \zeta_1)\right\}$$

is the region in which all the solutions of (2.2.1) with $e(t) = 0$, that start in $\mathbb{R}^3_+$, are confined. Therefore, it is the region of attraction in this situation. This completes the theorem. □

Remark: If the concentration of toxicant in resources is zero (i.e. $\alpha = 0$), then eventually the toxicant would be removed and the population would recover to its former level. □

If $e(t) = 0$, then the (2.2.1) has two non-negative equilibria in XEL-space:

$$P_0\left(0, 0, \frac{b}{\zeta_1}\right) \text{ and } P_1\left(\frac{K_0 g(b/\zeta_1)}{g_0}, 0, \frac{b}{\zeta_1}\right), \text{ where } b = \frac{d_1\alpha\beta}{a_1}. \tag{2.2.5}$$

The variational matrices of (2.2.1) at P_0 and P_1 are, respectively, as follows:

$$M_0 = \begin{bmatrix} g_0 & 0 & 0 \\ 0 & -\zeta_0 & 0 \\ 0 & a_1 & -\zeta_1 \end{bmatrix}, \; M_1 = \begin{bmatrix} -g(b/\zeta_1) & \frac{g^2(b/\zeta_1)K'(0)}{g_0} & \frac{g'(b/\zeta_1)K_0 g(b/\zeta_1)}{g_0} \\ 0 & -\zeta_0 & 0 \\ 0 & a_1 & -\zeta_1 \end{bmatrix}.$$

It is evident that P_0 is unstable (hyperbolic saddle) in the direction orthogonal to EL-plane and P_1 is locally asymptotically stable.

Situation II. Nonzero constant exogenous input: $e(t) = Q > 0$.

Let us state a lemma, which establishes a region of attraction for the system (2.2.1) with $e(t) = Q > 0$.

Lemma 2.1 *The set*

$$B = \left\{(X, E, L) : 0 < X \le K_0, \; 0 < E + L \le \frac{Q+b}{\zeta}, \; \zeta = \min(\zeta_0 - a_1, \zeta_1)\right\}$$

attracts all solutions of (2.2.1) (with $e(t) = Q > 0$) initially in the positive quadrant.

The proof is similar to the proof of Theorem 2.1. □

System (2.2.1) with $e(t) = Q > 0$ has the following two non-negative equilibria:

$$P_2\left(0, \frac{Q}{\zeta_0}, \frac{a_1 Q + b\zeta_0}{\zeta_0\zeta_1}\right) \text{ and } \overline{P}(\overline{X}, \overline{E}, \overline{L}),$$

where

$$\overline{X} = \frac{g(\overline{L})K(\overline{E})}{g_0}, \; \overline{E} = \frac{Q}{\zeta_0} \text{ and } \overline{L} = \frac{1}{\zeta_1}\left(a_1\frac{Q}{\zeta_0} + b\right). \tag{2.2.6}$$

The variational matrices at P_2 and $\overline{P}$ are, respectively,

$$M_2 = \begin{bmatrix} g_0 & 0 & 0 \\ 0 & -\zeta_0 & 0 \\ 0 & a_1 & -\zeta_1 \end{bmatrix}, \quad \overline{M} = \begin{bmatrix} -g(\overline{L}) & \frac{g^2(\overline{L})K'(\overline{E})}{g_0} & \frac{g(\overline{L})K(\overline{E})g'(\overline{L})}{g_0} \\ 0 & -\zeta_0 & 0 \\ 0 & a_1 & -\zeta_1 \end{bmatrix}.$$

Therefore, P_2 is unstable in the direction orthogonal to EL-plane and $\overline{P}$ is locally asymptotically stable (as in Situation I).

$\overline{P}$ is not always globally asymptotically stable. In the following, we are able to state conditions which guarantee the global stability of $\overline{P}$.

Theorem 2.2 *In addition to assumptions (2.2.2)–(2.2.4), let $g(L)$ and $K(E)$ satisfy in B:*

$$K_m \le K(E) \le K_0,\ 0 \le -K'(E) \le \kappa,\ 0 \le -g'(L) \le \gamma, \tag{2.2.7}$$

for some positive constants K_m, κ, γ. If the following inequalities hold:

$$\left[\frac{g_0 K_0 \kappa}{K_m^2}\right]^2 < \frac{g_0 \zeta_0}{K(\overline{E})}, \tag{2.2.8a}$$

$$\gamma^2 < \frac{g_0 \zeta_1}{K(\overline{E})}, \tag{2.2.8b}$$

$$a_1^2 < \zeta_0 \zeta_1, \tag{2.2.8c}$$

then $\overline{P}$ is globally asymptotically stable in B.

Proof Let us consider the following positive definite function about $\overline{P}$:

$$\Gamma(X, E, L) = X - \overline{X} - \overline{X} \ln \frac{X}{\overline{X}} + \frac{1}{2}(E - \overline{E})^2 + \frac{1}{2}(L - \overline{L})^2.$$

Differentiating Γ with respect to t along the solution of system (2.2.1) with $e(t) = Q > 0$, we get (after some simplifications):

$$\begin{aligned} \dot{\Gamma} = & -a_{11}(X - \overline{X})^2 - a_{22}(E - \overline{E})^2 - a_{33}(L - \overline{L})^2 + a_{12}(X - \overline{X})(E - \overline{E}) \\ & + a_{13}(X - \overline{X})(L - \overline{L}) + a_{23}(L - \overline{L})(E - \overline{E}), \end{aligned} \tag{2.2.9}$$

where

$$\left.\begin{aligned} & a_{11} = \frac{g_0}{K(\overline{E})}, \qquad a_{22} = \zeta_0, \qquad a_{33} = \zeta_1, \\ & a_{12} = -g_0 X \Upsilon_1(E),\ a_{13} = \Upsilon_2(L),\ a_{23} = a_1, \end{aligned}\right\} \tag{2.2.10a}$$

$$\Upsilon_1(E) = \begin{bmatrix} \left\{\dfrac{1}{K(E)} - \dfrac{1}{K(\overline{E})}\right\}/(E - \overline{E}), \; E \neq \overline{E}, \\ -\dfrac{K'(\overline{E})}{K(\overline{E})^2}, \qquad\qquad E = \overline{E}, \end{bmatrix} \tag{2.2.10b}$$

$$\Upsilon_2(L) = \begin{bmatrix} \{g(L) - g(\overline{L})\}(L - \overline{L}), \; L \neq \overline{L}, \\ g'(\overline{L}), \qquad\qquad L = \overline{L}. \end{bmatrix} \tag{2.2.10c}$$

Using (2.2.7), (2.2.10b, 2.2.10c) and the mean value theorem:

$$|\Upsilon_1(E)| \leq \frac{\kappa}{K_m^2}, \; |\Upsilon_2(L)| \leq \gamma. \tag{2.2.11}$$

Sufficient conditions for $\dot{\Gamma}$ to be negative definite are as follows:

$$a_{12}^2 - a_{11}a_{22} < 0, \; a_{13}^2 - a_{11}a_{33} < 0, \; a_{23}^2 - a_{22}a_{33} < 0. \tag{2.2.12}$$

It is evident that if the conditions (2.2.8a, 2.2.8b, 2.2.8c) hold then (2.2.12) will hold automatically. Therefore, Γ is a Lyapunov function with respect to $\overline{P}$ whose domain is contained within the region of attraction B. Hence, the theorem is proved. □

Remark: If Q and a_1 are small, then the possibility of satisfying conditions (2.2.8a)–(2.2.8c) increases. This indicates that if the rate of emission of toxicant into the environment and the population's rate of toxicant uptake from the environment both are small, then the stability of the dynamical system is more plausible. □

This theorem establishes that the underlying dynamical system settles down to a steady state of population (at a lower carrying capacity) and toxicant (at a level determined by influx and washout), provided inequalities (2.2.8a) to (2.2.8c) hold good.

Situation III. Rapidly fluctuating random exogenous input

In this situation, it is assumed that the exogenous input of toxicant into the environment ($= e(t)$) fluctuates randomly with respect to time and these fluctuations are rapid. The dynamical behaviour of system (2.2.1) under this situation will be considered within the framework of the following mathematical model:

$$\frac{\mathrm{d}}{\mathrm{d}t}x(t) = F(x(t), \; \xi(t/\epsilon)), \tag{2.2.13a}$$

where

$$x(t) = \begin{bmatrix} X(t) \\ E(t) \\ L(t) \end{bmatrix}, \quad F = \begin{bmatrix} g(L)X(t) - \frac{g_0 X^2(t)}{K(E)} \\ -\zeta_0 E + Q + \xi(t/\epsilon) \\ -\zeta_1 L + a_1 E + b \end{bmatrix},$$

where $0 < \epsilon << 1$ is a small, non-random parameter having the following meaning: as the *natural* time t changes by a typical amount (say) Δt, $\xi(t/\epsilon)$ fluctuates considerably, because it experiences an elapsed time $\Delta\tau = \frac{\Delta t}{\epsilon}$ which is large for small enough ϵ. The perturbed term $\xi(\tau)$, $\tau = t/\epsilon$, is a *coloured noise* or *Ornstein–Uhlenbeck process* characterized by:

$$\langle \xi(\tau) \rangle = 0, \quad \text{and} \quad \langle \xi(\tau_1)\xi(\tau_2) \rangle = \sigma\rho \exp(-\rho|\tau_1 - \tau_2|), \tag{2.2.13b}$$

where $\sigma, \rho^{-1} > 0$ are, respectively, the intensity, and the correlation time of the *coloured noise* $\xi(\tau)$ and $\langle\cdot\rangle$ represents the average over the ensemble of the stochastic process under consideration.

Perturbation approximation and non-equilibrium fluctuation

Let us use the following two-term perturbation approximation to $x(t)$ (White 1977):

$$x(t) \sim x^0(t) + \sqrt{\epsilon} Y^0(t). \tag{2.2.14}$$

The first approximation:

$$x^0(t) = \begin{bmatrix} X^0(t) \\ E^0(t) \\ L^0(t) \end{bmatrix},$$

satisfies

$$\frac{\mathrm{d}}{\mathrm{d}t} x^0(t) = \overline{F}(x^0(t)), \tag{2.2.15}$$

where

$$\overline{F}(x^0(t)) = \lim_{p\to\infty} \frac{1}{p} \int_0^p \langle F(x^0(t), \xi(\tau)) \rangle \mathrm{d}\tau$$

$$= \begin{bmatrix} g(U^0)X^0(t) - \frac{g_0 X^{0^2}(t)}{K(E^0)} \\ -\zeta_0 E^0(t) + Q \\ -\zeta_1 L^0(t) + a_1 E^0(t) + b \end{bmatrix}.$$

These are nothing but the equations of Situation II, and so this system has a unique non-trivial equilibrium at $\overline{P}(\overline{X}, \overline{E}, \overline{L})$ given by (2.2.6).

We assume that the system is at the initial time $t = 0$ at the point $\overline{P}$; therefore, $x^0(t) = \overline{P}$ for $t > 0$ ($\because$ $\overline{P}$ is locally asymptotically stable).

Here,

$$Y^0(t) = \begin{bmatrix} Y_1^0(t) \\ Y_2^0(t) \\ Y_3^0(t) \end{bmatrix}$$

is a Gaussian random process which satisfies the following linear equation:

$$\frac{\mathrm{d}}{\mathrm{d}t}Y^0(t) = \Theta Y^0(t) + \Psi(t), \tag{2.2.16}$$

where

$$Y^0(0) = O, \quad \Theta = \frac{\partial \overline{F}}{\partial x}(\overline{P}) = \begin{bmatrix} -g(\overline{L}) & \frac{g^2(\overline{L})K'(\overline{E})}{g_0} & \frac{g(\overline{L})K(\overline{E})g'(\overline{L})}{g_0} \\ 0 & -\zeta_0 & 0 \\ 0 & a_1 & -\zeta_1 \end{bmatrix}, \tag{2.2.17}$$

$$\langle \Psi(t) \rangle = O, \ \langle \Psi(t)\Psi^{\mathrm{T}}(t') \rangle = A\delta(t-t'), \tag{2.2.18}$$

and

$$A = \lim_{p\to\infty} \frac{1}{p} \int_0^p \int_0^p \langle [F(\bar{x}, \xi(\tau_1)) - \langle F(\bar{x}, \xi(\tau_1)) \rangle] [F(\bar{x}, \xi(\tau_2)) - \langle F(\bar{x}, \xi(\tau_2)) \rangle]^{\mathrm{T}} \rangle \mathrm{d}\tau_1 \mathrm{d}\tau_2 = \begin{bmatrix} 0 & 0 & 0 \\ 0 & 2\sigma & 0 \\ 0 & 0 & 0 \end{bmatrix}. \tag{2.2.19}$$

Here, $\delta(t)$ represents the Dirac delta function defined by

$$\begin{cases} \delta(t) = 0, \ \text{for } t \neq 0, \\ \int_{-\infty}^{\infty} \delta(t)\mathrm{d}t = 1. \end{cases}$$

The solution of (2.2.16) is as follows:

$$Y^0(t) = Y(t) \int_0^t Y^{-1}(s)\Psi(s)\mathrm{d}s, \tag{2.2.20}$$

where $Y(t)$ satisfies the following linear equation:

$$\frac{\mathrm{d}}{\mathrm{d}t}Y(t) = \Theta Y(t), \ Y(0) = I. \tag{2.2.21}$$

Therefore,

$$\langle Y^0(t) \rangle = O \ (\because \langle \Psi(s) \rangle = O). \tag{2.2.22}$$

The solution of (2.2.21) is as follows:

$$Y(t) = \begin{bmatrix} e^{-mt} & \left(n + \frac{qa_1}{\zeta_1-\zeta_0}\right)\left(\frac{e^{-mt}-e^{-\zeta_0 t}}{m-\zeta_0}\right) - \frac{qa_1(e^{-mt}-e^{-\zeta_1 t})}{(\zeta_1-\zeta_0)(m-\zeta_1)} & \frac{q(e^{-mt}-e^{-\zeta_1 t})}{m-\zeta_1} \\ 0 & e^{-\zeta_0 t} & 0 \\ 0 & \frac{a_1(e^{-\zeta_0 t}-e^{-\zeta_1 t})}{(\zeta_1-\zeta_0)} & e^{-\zeta_1 t} \end{bmatrix}, \quad (2.2.23)$$

where

$$m = g(\overline{L}), \quad n = -\frac{g^2(\overline{L})K'(\overline{E})}{g_0}, \quad q = -\frac{g(\overline{L})K(\overline{E})g'(\overline{L})}{g_0}. \quad (2.2.24)$$

The expression of the strength of the fluctuation $D(t)$, the covariance at one instant of time t, is given by:

$$D(t) = \langle Y^0(t)Y^{0^{\mathrm{T}}}(t)\rangle = Y(t)\left[\int_0^t Y^{-1}(s)AY^{-1^{\mathrm{T}}}(s)\mathrm{d}s\right]Y^T(t) = [D_{ij}(t)] \text{ (say)}. \quad (2.2.25)$$

After some simplifications:

$$\begin{aligned} D_{11}(t) = \mathrm{var}[X(t)] = \sigma \Bigg[& \frac{(\Lambda_1+\Lambda_2)^2}{a}\mathrm{e}^{2at} - \frac{\Lambda_2^2}{\zeta_0}\mathrm{e}^{-2\zeta_0 t} - \frac{\Lambda_1^2}{\zeta_1}\mathrm{e}^{-2\zeta_1 t} \\ & + \frac{4\Lambda_2(\Lambda_1+\Lambda_2)}{\zeta_0 - a}\mathrm{e}^{(a-\zeta_0)t} + \frac{4\Lambda_1(\Lambda_1+\Lambda_2)}{\zeta_1 - a}\mathrm{e}^{(a-\zeta_1)t} \\ & - \frac{4\Lambda_1\Lambda_2}{\zeta_0+\zeta_1}\mathrm{e}^{-(\zeta_0+\zeta_1)t} - \frac{(\Lambda_1+\Lambda_2)^2}{a} + \frac{\Lambda_2^2}{\zeta_0} + \frac{\Lambda_1^2}{\zeta_1} \\ & - \frac{4\Lambda_2(\Lambda_1+\Lambda_2)}{\zeta_0 - a} - \frac{4\Lambda_1(\Lambda_1+\Lambda_2)}{\zeta_1 - a} + \frac{4\Lambda_1\Lambda_2}{\zeta_0+\zeta_1}\Bigg], \end{aligned} \quad (2.2.26)$$

where

$$a = -g(\overline{L}), \quad \Lambda_1 = \frac{g(\overline{L})K(\overline{E})g'(\overline{L})a_1}{g_0(\zeta_1-\zeta_0)(a+\zeta_1)},$$

and

$$\Lambda_2 = -\frac{1}{a+\zeta_0}\left\{\frac{g^2(\overline{L})K'(\overline{E})}{g_0} + \frac{g(\overline{L})K(\overline{E})g'(\overline{L})a_1}{g_0(\zeta_1-\zeta_0)}\right\}.$$

In special case, taking the limit as $\zeta_1 \to \zeta_0$:

$$\lim_{\zeta_1\to\zeta_0} D_{11}(\infty) = \frac{\sigma}{m\zeta_0(\zeta_0+m)}\left(n + \frac{qa_1}{\zeta_0 - m}\right)^2, \quad (2.2.27)$$

where

$$m = g(\overline{L}),\ \ n = -\frac{g^2(\overline{L})K'(\overline{E})}{g_0},\ \ q = -\frac{g(\overline{L})K(\overline{E})g'(\overline{L})}{g_0}.$$

For constant exogenous input of toxicant into the environment: the eigen values of the variational matrix of the underlying system are $-\zeta_0, -\zeta_1, -m$. Therefore, the deterministic local stability criterion is fulfilled (since $\zeta_0, \zeta_1, m > 0$). In the stochastic (fluctuating) emission of toxicant by external sources into the environment: the linear (local) stability of the underlying single-species population is again characterized by ζ_0, m (provided $\zeta_1 \to \zeta_0$), but in this situation, it is no longer enough that $\Xi = \min(\zeta_0, m) > 0$, because if $\sigma >> \Xi$, population exhibits large fluctuations leading to rapid extinction (using (2.2.27)). It is remembered that σ is the intensity of the environmental noise. When σ and Ξ are commensurate: the system is likely to undergo significant fluctuations, even though they persist for long time. For the situations $\sigma << \Xi$: population fluctuations are relatively small and the emission of toxicant into the environment is effectively deterministic (non-fluctuating). These results agree with May (1973).

2.3 Dynamical Model II of a Single-Species System in a Polluted Environment

In this article, we have developed a single-species population model in polluted environment. In Sect. 2.3.2, we present a brief sketch of the construction of the model. Boundedness and positivity analysis of the solutions are shown which implies that the system is ecologically well-behaved. In the next section, we have discussed the existence and stability analysis of the model at various equilibrium points under zero exogenous input and nonzero exogenous input. Then, we obtain necessary conditions for the existence of interior equilibrium P^* and local and global stability of the system at P^*. The analysis of double-delayed model is described in Sect. 2.3.4. The stability of the model with double delays is investigated by the Nyquist criteria. By choosing one of the delays as a bifurcation parameter, the model is found to undergo a Hopf bifurcation. Some of the important analytic results are numerically verified by using MATLAB in Sect. 2.3.5.

2.3.1 Model Construction

Now, we shall discuss about the construction of our model based on the procedure developed by Buonomo et al. (1999). Let B be a bounded domain in $\mathbb{R}^i (i = 1, 2, 3)$ with smooth boundary ∂B. Let V be an arbitrary domain contained in B. The amount of population living in V at time t is denoted by $\mathbb{P}_V(t)$, the total amount of a given (external) toxicant in V at time t is denoted by $E_V(t)$, and the total amount of the toxicant stored inside the bodies of the organisms living in V at time t is denoted by $T_V(t)$. We assume that there exist three smooth functions $\rho(x,t), c(x,t), e(x,t)$ defined in $B \times (0,T]$, $(T \leq +\infty)$ such that for any V in B it results:

$$\mathbb{P}_V(t) = \int \rho(x,t)\mathrm{d}x; \quad E_V(t) = \int e(x,t)\mathrm{d}x; \quad T_V(t) = \int c(x,t)\rho(x,t)\mathrm{d}x.$$

Here, $\mathbb{P}_V$ and E_V are defined through respective densities, while T_V is defined through the density of $\mathbb{P}_V$ weighted by a concentration c. Let us derive the equations for ρ, c, e by balance arguments. For this purpose, assume that the population dynamics is governed by:

$$\dot{\mathbb{P}}_V = \int_V [\rho f(\rho,c) - \mathrm{div} J_\rho]\mathrm{d}x,$$

where J_ρ is the flux of the population through the boundary of V. The growth rate $f(\rho,c)$ of the population is a smooth function given by the difference between the birth rate $b(\rho)$ and the death rate $\bar{d}(c)$.
For the external toxicant, we have

$$\dot{E}_V = \int_V [-ke\rho + \{r + \bar{d}(c)\}c\rho - he + \Gamma - \mathrm{div} J_e]\mathrm{d}x.$$

The term $-ke\rho$ represents the uptake of the toxicant from the environment by population; $rc\rho$ is the toxicant input to the environment from the population due to egestion; $\bar{d}(c)c\rho$ represents the amount of toxicant stored by the living organisms which die at time t; he denotes losses from the environment outside the system; Γ is the exogenous toxicant input rate which is assumed to be a smooth bounded non-negative function of x and t.
For the internal toxicant:

$$\dot{T}_V = \int_V [ke\rho - \{r + m + \bar{d}(c)\}c\rho - \mathrm{div} J_T]\mathrm{d}x.$$

The term $-mc\rho$ represents metabolization processes and other losses, J_e and J_T denote, respectively, the flux of external and internal toxicant through the boundary of V.
Under the assumption of a *Fickian diffusion mechanism* for the population and the external toxicant:

$$J_\rho = -\delta_\rho \nabla \rho, \quad J_e = -\delta_e \nabla e,$$

where δ_ρ and δ_e are the diffusion coefficients (positive constants). As for the flux J_T, since the internal toxicant is in some sense drifted by the random walk of the living population, it seems rather natural to assume $J_T = cJ_\rho = -\delta_\rho c \nabla \rho$. Concerning the

growth rate of the population, we assume that the birth rate is $b(\rho) = b_0 - g\rho$ and the death rate is $\bar{d}(c) = d_0 + \gamma c$, where b_0, d_0 and γ are positive constants. Here, g is assumed to be a non-negative constant: in the absence of toxicant, a *malthusian* ($g = 0$) or a *logistic growth rate* ($g > 0$).
Because of the arbitrariness of V and using standard regularity assumptions, the following system of partial differential equations arises:

$$\begin{aligned} &\frac{\partial \rho}{\partial t} - \delta_\rho \Delta \rho = \rho(b_0 - d_0 - \gamma c - g\rho) \\ &\frac{\partial c\rho}{\partial t} - \delta_\rho c \Delta \rho - \delta_\rho \nabla c \cdot \nabla \rho = ke\rho - (r + m + d_0 + \gamma c)c\rho \\ &\frac{\partial e}{\partial t} - \delta_e \Delta e = -ke\rho + (r + d_0 + \gamma c)c\rho - he + \Gamma \end{aligned} \tag{I}$$

in $B \times (0, T]$.
The second equation in (I) can be written as

$$\rho \frac{\partial c}{\partial t} + c \frac{\partial \rho}{\partial t} - \delta_\rho c \Delta \rho - \delta_\rho \nabla c \cdot \nabla \rho = ke\rho - (r + m + d_0 + \gamma c)c\rho \tag{II}$$

Thus, system (I) is equivalent to

$$\begin{aligned} &\frac{\partial \rho}{\partial t} - \delta_\rho \Delta \rho = \rho(b_0 - d_0 - \gamma c - g\rho) \\ &\rho \frac{\partial c}{\partial t} - \delta_\rho \nabla c \cdot \nabla \rho = ke\rho - (r + m + b_0 - g\rho)c\rho \\ &\frac{\partial e}{\partial t} - \delta_e \Delta e = -ke\rho + (r + d_0 + \gamma c)c\rho - he + \Gamma \end{aligned} \tag{III}$$

in $B \times (0, T]$.
We further suppose no-flux through the boundary ∂B. Precisely, we append to system (I) homogeneous Neumann boundary conditions:

$$\frac{\partial \rho}{\partial n} = \frac{\partial e}{\partial n} = 0, \quad \text{on } \partial B \times (0, T].$$

Finally, we set initial data as:

$$\rho(x, 0) = \rho_0(x) > 0; \;\; c(x, 0) = c_0(x) \geq 0; \;\; e(x, 0) = e_0(x) \geq 0; \;\; x \in \overline{B}.$$

If the initial data are constants, then the system (III) reduces to the following system of ordinary differential equations:

$$\begin{aligned}
\frac{d\rho}{dt} &= \rho(b_0 - d_0 - \gamma c - g\rho), \\
\frac{dc}{dt} &= ke - (r + m + b_0 - g\rho)c, \qquad \text{(IV)} \\
\frac{de}{dt} &= -ke\rho + (r + d_0 + \gamma c)c\rho - he + \Gamma.
\end{aligned}$$

Basic mathematical model

Here, we assume:

$\rho(t)$: concentration of the population biomass,

$c(t)$: concentration of the toxicant in the population,

$e(t)$: concentration of the toxicant in the environment.

The model satisfies the following assumptions:

(**A1**) There is a given toxicant in the environment and the living organisms absorb into their bodies part of this toxicant so that the dynamics of the population is affected by the toxicant.

(**A2**) For the growth rate of population, we assume that the birth rate is $b_0 - g\rho(t)$ and the death rate is $d_0 + \gamma c(t)$, where b_0, g, d_0, γ are assumed to be positive constants.

We consider the following model:

$$\begin{aligned}
\frac{d\rho}{dt} &= \rho(b_0 - d_0 - \gamma c - g\rho), \\
\frac{dc}{dt} &= ke - (r + m + b_0 - g\rho)c, \\
\frac{de}{dt} &= -ke\rho + (r + d_0 + \gamma c)c\rho - he + \Gamma(t),
\end{aligned} \tag{2.3.1}$$

with initial data $\rho(0) > 0, c(0) \geq 0, e(0) \geq 0$.

Here,

k: depletion rate of toxicant in the environment due to its intake made by the population,

r: depletion rate of toxicant in the population due to egestion,

m: depletion rate of toxicant in the population due to metabolization process,

h: depletion rate of toxicant in the environment due to other causes,

$\Gamma(t)$: exogenous toxicant input rate which is assumed to be a smooth bounded non-negative function of t.

We can see that, if $b_0 - d_0 - \gamma c(t) \leq 0$, then $\rho(t)$ will be going to extinct. So we suppose,

$$\alpha = b_0 - d_0 > 0 \text{ and } b_0 - d_0 - \gamma c(t) > 0 \Rightarrow c(t) < \frac{b_0 - d_0}{\gamma}, \forall t \geq 0 \tag{2.3.2}$$

The model stated in (2.3.1) has nine parameters, which make the analysis difficult. To reduce the number of parameters and to determine which combinations of parameters control the behaviour of the system, let us choose

$$X = g\rho, Y = \gamma c, Z = k\gamma e$$

So, system (2.3.1) becomes:

$$\begin{aligned}
\frac{\mathrm{d}X}{\mathrm{d}t} &= X(\alpha - Y - X), \\
\frac{\mathrm{d}Y}{\mathrm{d}t} &= Z - (\beta - X)Y, \\
\frac{\mathrm{d}Z}{\mathrm{d}t} &= -aXZ + a(d + Y)XY - hZ + u(t),
\end{aligned} \tag{2.3.3}$$

with initial data:

$$X(0) > 0, Y(0) \geq 0, Z(0) \geq 0, \tag{2.3.4}$$

where

$\alpha = b_0 - d_0, \beta = r + m + b_0, a = \dfrac{k}{g}, d = r + d_0, u(t) = k\gamma\Gamma(t).$

Therefore, from (2.3.2), we have

$$Y(t) < \alpha, \qquad \forall t \geq 0 \;\; [\because Y(t) = \gamma c(t)] \tag{2.3.5}$$

Theorem 2.3 *Each component of the solution of system (2.3.3) subject to (2.3.4) is non-negative and bounded for all* $t > 0$.

Proof Since the right-hand side of system (2.3.1) is completely continuous and locally Lipschitzian on C, the solution $(X(t), Y(t), Z(t))$ of (2.3.3) with initial conditions (2.3.4) exists and is unique on $[0, \zeta)$, where $0 < \zeta \leq +\infty$ (Hale 1977). Now, from the first equation of system (2.3.3), we have

$$X(t) = X(0) \exp \int_0^t \{\alpha - Y(s) - X(s)\} \mathrm{d}s > 0, \; \forall\, t \geq 0.$$

Next, we show that $Y(t) \geq 0$ for all $t \in [0, \zeta)$, where $0 < \zeta \leq +\infty$. Otherwise, there exists a $t_1 \in [0, \zeta)$ such that $Y(t_1) = 0, \;\; \dot{Y}(t_1) < 0$ and $Y(t) \geq 0$ for all $t \in [0, t_1]$. Hence, it must have $Z(t) \geq 0$ for all $t \in [0, t_1]$. If this statement is not true, then there exists a $t_2 \in [0, t_1)$ such that $Z(t_2) = 0, \;\; \dot{Z}(t_2) < 0$ and $Z(t) \geq 0$ for all $t \in [0, t_2]$. From the third equation of (2.3.3), we have:

$$\dot{Z}(t_2) = a\,(d + Y(t_2))\, X(t_2) Y(t_2) + u(t_2) \geq 0,$$

which is a contradiction with $\dot{Z}(t_2) < 0$. So, $Z(t) \geq 0$ for all $t \in [0, t_1]$. Now from the second equation of (2.3.3), we have:

$$\dot{Y}(t_1) = Z(t_1) \geq 0, \tag{2.3.6}$$

which is a contradiction with $\dot{Y}(t_1) < 0$. So $Y(t) \geq 0,\ \forall\, t \geq 0$ and hence $Z(t) \geq 0,\ \forall\, t \geq 0$. Therefore,

$$X(t) > 0,\, Y(t) \geq 0,\, Z(t) \geq 0,\ \forall\, t \geq 0.$$

From the first equation of (2.3.3) and (2.3.6), we have:

$$\frac{\mathrm{d}X}{\mathrm{d}t} = X\,(\alpha - Y - X) \leq \alpha X \left(1 - \frac{X}{\alpha}\right)$$

Therefore, by a standard comparison theorem, we have,

$$\limsup_{t\to\infty} X\,(t) \leq \alpha. \tag{2.3.7}$$

From the third equation of (2.3.3), (2.3.5), (2.3.6) and (2.3.7), we have:

$$\begin{aligned} \frac{\mathrm{d}Z}{\mathrm{d}t} &< a\,(d+\alpha)\,\alpha X - hZ + u^M \Rightarrow \limsup_{t\to\infty} Z(t) \\ &\leq \frac{a(d+\alpha)\alpha^2 + u^M}{h}, \text{ where } u^M = \limsup_{t\geq 0} u(t). \end{aligned} \tag{2.3.8}$$

From (2.3.5)–(2.3.8), we conclude that each component of the solution of system (2.3.3) subject to (2.3.4) is non-negative and bounded for all $t > 0$.

This completes the proof.

2.3.2 *Routh–Hurwitz Criterion for Local Stability*

The Routh–Hurwitz criterion can be used to determine whether an equilibrium is stable without having to calculate the eigenvalues explicitly. If the system is of nth order, then the characteristic equation can be taken in the general form:

$$\lambda^n + a_1\lambda^{n-1} + \cdots + a_n = 0,$$

where the coefficients $a_i, i = 1, 2, \ldots, n$ are all real. We tacitly assume that $a_n \neq 0$ since otherwise $\lambda = 0$ is a solution and the polynomial is then of order $n - 1$. An equilibrium will be locally stable in a continuous time model (i.e. all eigenvalues will have negative real parts) if and only if the following determinants are all positive:

$$A_1 = |a_1| > 0, \quad A_2 = \begin{vmatrix} a_1 & a_3 \\ 1 & a_2 \end{vmatrix} > 0, \quad A_3 = \begin{vmatrix} a_1 & a_3 & a_5 \\ 1 & a_2 & a_4 \\ 0 & a_1 & a_3 \end{vmatrix} > 0, \ldots,$$

$$A_k = \begin{vmatrix} a_1 & a_3 & a_5 & \cdots\cdots \\ 1 & a_2 & a_4 & \cdots\cdots \\ 0 & a_1 & a_3 & \cdots\cdots \\ 0 & 1 & a_2 & \cdots\cdots \\ \cdots & \cdots & \cdots & \cdots \\ 0 & 0 & \cdots\cdots & a_k \end{vmatrix} > 0, \quad k = 1, 2, \ldots, n; \text{ where } a_j = 0 \text{ if } j > n.$$

These conditions are derived, using complex variable methods, in standard texts on the theory of dynamical systems. As an example, for the characteristic equation:

$$\lambda^3 + a_1\lambda^2 + a_2\lambda + a_3 = 0,$$

the conditions for $\mathrm{Re}(\lambda) < 0$ are $a_1 > 0,\ a_3 > 0,\ a_1a_2 - a_3 > 0$.

2.3.3 Hopf Bifurcation Theorem

The conditions for occurrence of a simple Hopf bifurcation in a general three-species population model are stated in the following theorem:

Theorem 2.4 *Suppose that the interior equilibrium depends smoothly on some parameter ξ in an open interval $R_1 \subseteq R$. If $\exists\, \xi^* \in R_1$ such that*

(a) a simple pair of complex eigenvalues of the variational matrix at the interior equilibrium point exists, say $p(\xi) \pm iq(\xi)$ such that they becomes purely imaginary at $\xi = \xi^*$, whereas the other eigenvalue remains real and negative; and

(b) $\left[\dfrac{dp}{d\xi}\right]_{\xi=\xi^*} \neq 0,$

then at ξ^* a simple Hopf bifurcation occurs (i.e. change of stability behaviour with creation of a limit cycle).

Liu (1994) derived a criterion of Hopf bifurcation without using the eigenvalues of the variational matrix at the interior equilibrium point which is stated below for the purpose of the present study:

Liu's criterion If the characteristic equation at the equilibrium point is as follows: $\lambda^3 + a_1(\xi)\lambda^2 + a_2(\xi)\lambda + a_3(\xi) = 0$, where $a_1(\xi)$, $D(\xi) = a_1(\xi)a_2(\xi) - a_3(\xi)$, $a_3(\xi)$ are smooth functions of ξ in an open interval about $\xi^*(\in R)$ such that

(i) $a_1(\xi^*) > 0,\ D(\xi^*) = 0,\ a_3(\xi^*) > 0,$

(ii) $\left[\dfrac{\mathrm{d}D}{\mathrm{d}\xi}\right]_{\xi=\xi^*} \neq 0,$

then a Hopf bifurcation occurs at $\xi = \xi^*$.

2.3.4 *Stability Behaviour of the Model*

Case I: Zero exogenous input $(u(t) = 0)$

Theorem 2.5 *If $u(t) = 0$, then the model (2.3.3) has non-negative equilibria $P_0(0, 0, 0)$ which is unstable and $P_1(\alpha, 0, 0)$ which is locally asymptotically stable. The interior equilibrium $P^*(X^*, Y^*, Z^*)$ is not feasible.*

Proof The variational matrix of system (2.3.3) at P_0 is

$$\Lambda(P_0) = \begin{bmatrix} \alpha & 0 & 0 \\ 0 & -\beta & 1 \\ 0 & 0 & -h \end{bmatrix}$$

The eigen values are $\alpha, -\beta, -h$. So, it is obvious that, P_0 is unstable (hyperbolic saddle).
Now, the variational matrix of system (2.3.3) at P_1 is

$$\Lambda(P_1) = \begin{bmatrix} -\alpha & -\alpha & 0 \\ 0 & -(\beta - \alpha) & 1 \\ 0 & ad\alpha & -(a\alpha + h) \end{bmatrix}$$

The characteristic equation of $\Lambda(P_1)$ is $(\alpha + \lambda)\left(\lambda^2 + C_1\lambda + C_2\right) = 0$,
where $C_1 = a\alpha + h + \beta - \alpha = a\alpha + h + d_0 + r + m > 0$, and $C_2 = (a\alpha + h)(\beta - \alpha) - a\alpha d = a\alpha(\beta - \alpha - d) + h(\beta - \alpha) = a\alpha m + h(d_0 + r + m) > 0$,
since $\alpha > 0$.
The eigen values are $\lambda_1 = -\alpha < 0$ and $\lambda_{2,3} = \frac{-C_1 \pm \sqrt{C_1^2 - 4C_2}}{2}$.
Since $C_1 > 0$, $C_2 > 0$; therefore, the signs of the real parts of λ_2, λ_3 are negative.
Hence, P_1 is locally asymptotically stable.
It is noted here that the other equilibrium point (X^*, Y^*, Z^*) is not feasible, since

$$X^* = \frac{h\beta}{h - am}, \quad Y^* = \frac{-h(r + m + d_0) - a\alpha m}{h - am}$$

are opposite in sign since $\alpha > 0$, by (2.3.2).
Hence, the theorem is proved.

Case II: Nonzero exogenous input $(u(t) = Q > 0)$
When $(u(t) = Q > 0)$, the model (2.3.3) has two non-negative equilibria, $P_2\left(0, \frac{Q}{h\beta}, \frac{Q}{h}\right)$ and $P^*(X^*, Y^*, Z^*)$. The variational matrix of system (2.3.3) at P_2 is given by

$$\Lambda(P_2) = \begin{bmatrix} \alpha - \frac{Q}{h\beta} & 0 & 0 \\ \frac{Q}{h\beta} & -\beta & 1 \\ -\frac{aQ}{h} + \frac{aQ}{h\beta}\left(d + \frac{Q}{h\beta}\right) & 0 & -h \end{bmatrix}$$

The characteristic equation of $\Lambda(P_2)$ is $\left(\alpha - \frac{Q}{h\beta} - \lambda\right)(\beta + \lambda)(h + \lambda) = 0$.
So, P_2 is asymptotically stable if and only if $h\alpha\beta < Q$ and if $h\alpha\beta > Q$, then P_2 becomes unstable.
The interior equilibrium point $P^*(X^*, Y^*, Z^*)$ of system (2.3.3) is given by

$$X^* = \frac{C_3 - h\beta + \alpha(am - h)}{2(am - h)}, \quad Y^* = \frac{\alpha(am - h) + h\beta - C_3}{2(am - h)},$$

$$Z^* = \frac{\{\alpha(am - h) + h\beta - C_3\}\{(am - h)(2\beta - \alpha) + h\beta - C_3\}}{4(am - h)^2}$$

where

$$C_3 = \sqrt{\{h\beta - \alpha(am - h)\}^2 + 4(am - h)(h\alpha\beta - Q)} \text{ and } m = (\beta - \alpha - d).$$

It can be proved that the unique interior equilibrium point $P^*(X^*, Y^*, Z^*)$ of system (2.3.3) exists if and only if the following two conditions are satisfied

$$\text{(i) } am > h \quad \text{and} \quad \text{(ii) } h\alpha\beta > Q$$

Summarizing the above analysis, we come to the following theorem:

Theorem 2.6 *If $u(t) = Q > 0$, then the equilibrium point $P_2(0, \frac{Q}{h\beta}, \frac{Q}{h})$ of the system (2.3.3) is locally asymptotically stable if and only if $h\alpha\beta < Q$ and the unique interior equilibrium point $P^*(X^*, Y^*, Z^*)$ of system (2.3.3) exists if and only if the following two conditions are satisfied*

$$\textit{(i) } am > h \quad \textit{and} \quad \textit{(ii) } h\alpha\beta > Q, \quad m = (\beta - \alpha - d)$$

Local stability of $P^*(X^*, Y^*, Z^*)$

The variational matrix at P^* is given by

$$\Lambda(P^*) = \begin{bmatrix} m_{11} & m_{12} & 0 \\ m_{21} & m_{22} & m_{23} \\ m_{31} & m_{32} & m_{33} \end{bmatrix}$$

where

$$m_{11} = -X^*, \quad m_{12} = -X^*, \quad m_{21} = Y^*, \quad m_{22} = -q + X^*, \quad m_{23} = 1,$$

$$m_{31} = -aZ^* + aY^*(d + Y^*), \quad m_{32} = aX^*(d + 2Y^*), \quad m_{33} = -(aX^* + h)$$

The characteristic equation is

$$\lambda^3 + \Upsilon_1\lambda^2 + \Upsilon_2\lambda + \Upsilon_3 = 0$$

where

$$\Upsilon_1 = -m_{11} - m_{22} - m_{33} = -\mathrm{tr}\left[\Lambda\left(P^*\right)\right]$$

$$\Upsilon_2 = m_{11}m_{22} + m_{11}m_{33} + m_{22}m_{33} - m_{23}m_{32} - m_{12}m_{21}$$

$$\Upsilon_3 = -\det\left[\Lambda\left(P^*\right)\right] = m_{11}m_{23}m_{32} + m_{12}m_{21}m_{33} - m_{11}m_{22}m_{33} - m_{12}m_{23}m_{31}$$

By the Routh–Hurwitz criterion, it follows that all eigenvalues of characteristic equation have negative real part if and only if

$$\Upsilon_1 > 0, \quad \Upsilon_3 > 0, \quad \Upsilon_1\Upsilon_2 - \Upsilon_3 > 0 \tag{2.3.9}$$

Theorem 2.7 *P^* is locally asymptotically stable if and only if the inequalities (2.3.9) are satisfied.*

Further, the conditions arising through Hopf bifurcation cannot be derived explicitly in terms of system parameters. Let us assume ξ stands for any parameter involved with the model system (2.3.3). If there exists a critical magnitude $\xi = \xi^*$ such that $\Upsilon_1(\xi^*) > 0$, $\Upsilon_3(\xi^*) > 0$, $(\Upsilon_1\Upsilon_2 - \Upsilon_3)_{\xi=\xi_*} = 0$ and $\left[\frac{\mathrm{d}}{\mathrm{d}\xi}(\Upsilon_1\Upsilon_2 - \Upsilon_3)\right]_{\xi=\xi^*} \neq 0$, then P^* undergoes a Hopf bifurcation at $\xi = \xi^*$ (Liu's criterion).

Global stability of P^*

P^* is not always globally asymptotically stable. The conditions which guarantee the global stability of P^* are stated in the following theorem:

Theorem 2.8 *Since $X(t)$, $Y(t)$, $Z(t)$ are bounded, there exist positive constants σ_i, Δ_i (i=1,2,3) such that $\sigma_1 \le X(t) \le \Delta_1$, $\sigma_2 \le Y(t) \le \Delta_2$, $\sigma_3 \le Z(t) \le \Delta_3$. If the following inequalities hold:*

$$\begin{aligned}
&(\Delta_2 - 1)^2 < \left(\beta - X^*\right) \\
&\{a\Delta_2(d + \Delta_2) - a\sigma_3\}^2 < \left(aX^* + h\right) \\
&\left(1 + adX^* + 2aX^*\Delta_2\right)^2 < \left(\beta - X^*\right)\left(h + aX^*\right),
\end{aligned} \tag{2.3.10}$$

then P^* is globally asymptotically stable.

Proof We consider the following positive definite function about P^* :

$$\Omega(X, Y, Z) = \left(X - X^* - X^* \ln\frac{X}{X^*}\right) + \frac{1}{2}\left(Y - Y^*\right)^2 + \frac{1}{2}\left(Z - Z^*\right)^2$$

Differentiating both sides with respect to t along the solution of (2.3.3), we get (after some simple calculations):

$$
\begin{aligned}
\frac{d\Omega}{dt} &= -\left(X - X^*\right)^2 + (Y-1)\left(X - X^*\right)\left(Y - Y^*\right) - \left(q - X^*\right)\left(Y - Y^*\right)^2 \\
&\quad + (aY(d+Y) - aZ)\left(X - X^*\right)\left(Z - Z^*\right) - \left(aX^* + h\right)\left(Z - Z^*\right)^2 \\
&\quad + \left(1 + adX^* + aX^*\left(Y + Y^*\right)\right)\left(Y - Y^*\right)\left(Z - Z^*\right) \\
&\leq -\left(X - X^*\right)^2 + (\Delta_2 - 1)\left(X - X^*\right)\left(Y - Y^*\right) - \left(\beta - X^*\right)\left(Y - Y^*\right)^2 \\
&\quad - \left(aX^* + h\right)\left(Z - Z^*\right)^2 + (a\Delta_2(d + \Delta_2) - a\sigma_3)\left(X - X^*\right)\left(Z - Z^*\right) \\
&\quad + \left(1 + adX^* + 2a\Delta_2 X^*\right)\left(Y - Y^*\right)\left(Z - Z^*\right) \\
&= -a_{11}\left(X - X^*\right)^2 - a_{22}\left(Y - Y^*\right)^2 - a_{33}\left(Z - Z^*\right)^2 + a_{12}\left(X - X^*\right)\left(Y - Y^*\right) \\
&\quad + a_{13}\left(X - X^*\right)\left(Z - Z^*\right) + a_{23}\left(Y - Y^*\right)\left(Z - Z^*\right)
\end{aligned}
$$

where

$$
a_{11} = 1,\ \ a_{22} = \beta - X^*,\ \ a_{33} = aX^* + h,\ \ a_{12} = \Delta_2 - 1,
$$

$$
a_{13} = \{a\Delta_2(d + \Delta_2) - a\sigma_3\},\ \ a_{23} = 1 + adX^* + 2a\Delta_2 X^*
$$

Now sufficient conditions for $\frac{d\Omega}{dt}$ to be negative definite are

$$
a_{12}^2 - a_{11}a_{22} < 0,\ \ a_{13}^2 - a_{11}a_{33} < 0,\ \ a_{22}^2 - a_{22}a_{33} < 0, \tag{2.3.11}
$$

i.e.

$$
\begin{aligned}
&(\Delta_2 - 1)^2 < \left(\beta - X^*\right) \\
&\{a\Delta_2(d + \Delta_2) - a\sigma_3\}^2 < \left(aX^* + h\right) \\
&\left(1 + adX^* + 2aX^*\Delta_2\right)^2 < \left(\beta - X^*\right)\left(h + aX^*\right).
\end{aligned}
$$

Hence, Ω is a Lyapunov function. This completes the proof. □

2.3.5 *Model with Double Delays*

In this section, we have discussed the model with double discrete time delays τ_1 and τ_2, where τ_1 represents the activation period or reaction time of the toxicant in the population biomass and τ_2 represents the activation period or reaction time of the toxicant in the population comes from the environment. The model is as follows:

$$
\begin{aligned}
\frac{dX}{dt} &= X\left(\alpha - Y(t - \tau_1) - X\right) \\
\frac{dY}{dt} &= Z(t - \tau_2) - (\beta - X)Y \\
\frac{dZ}{dt} &= -aXZ + a(d + Y)XY - hZ + Q
\end{aligned} \tag{2.3.12a}
$$

$$X(0) > 0; \quad Y(\theta_1) \geq 0, \quad \theta_1 \in [-\tau_1, 0]; \quad Z(\theta_2) \geq 0, \quad \theta_2 \in [-\tau_2, 0]. \qquad (2.3.12b)$$

Estimation of the length of delay to preserve stability

Now, we shall estimate the length of the delay which preserves the stability of the system (2.3.12a, 2.3.12b). The corresponding linearized system of (2.3.12a, 2.3.12b) about the interior equilibrium $P^*(X^*, Y^*, Z^*)$ is given by:

$$\begin{aligned}\frac{\mathrm{d}x_1}{\mathrm{d}t} &= a_{11}x_1 + a_{12}y_1(t-\tau_1)\\ \frac{\mathrm{d}y_1}{\mathrm{d}t} &= a_{21}x_1 + a_{22}y_1 + b_{23}z_1(t-\tau_2)\\ \frac{\mathrm{d}z_1}{\mathrm{d}t} &= a_{31}x_1 + a_{32}y_1 + a_{33}z_1\end{aligned} \qquad (2.3.13)$$

where

$$X(t) = x_1(t) + X^*, \quad Y(t) = y_1(t) + Y^*, \quad Z(t) = z_1(t) + Z^*.$$

Taking Laplace transform of the system (2.3.13), we get

$$\begin{aligned}(s - a_{11})\overline{x_1}(s) &= a_{12}\mathrm{e}^{-s\tau_1}\overline{y_1}(s) + a_{12}\mathrm{e}^{-s\tau_1}k_1(s) + x_1(0)\\ (s - a_{22})\overline{y_1}(s) &= a_{21}\overline{x_1}(s) + b_{23}\mathrm{e}^{-s\tau_2}\overline{z_1}(s) + b_{23}\mathrm{e}^{-s\tau_2}k_1(s) + y_1(0)\\ (s - a_{33})\overline{z_1}(s) &= a_{31}\overline{x_1}(s) + a_{32}\overline{y_1}(s) + z_1(0)\end{aligned} \qquad (2.3.14)$$

where

$$k_1(s) = \int_{-\tau_1}^{0} \mathrm{e}^{-st} y_1(t)\mathrm{d}t,$$

$$k_2(s) = \int_{-\tau_2}^{0} \mathrm{e}^{-st} z_1(t)\mathrm{d}t$$

and $\overline{x_1}(s), \overline{y_1}(s), \overline{z_1}(s)$ are the Laplace transforms of $x_1(t), y_1(t), z_1(t)$, respectively.

Now, using *Nyquist theorem*, it can be shown that the conditions for local asymptotic stability of $P^*(X^*, Y^*, Z^*)$ are given by (see Erbe et al. 1986):

$$\mathrm{Im}H(i\eta_0) > 0, \qquad (2.3.15)$$

$$\mathrm{Re}H(i\eta_0) = 0, \qquad (2.3.16)$$

where

$$H(s) = s^3 + D_1 s^2 + D_2 s + D_3 + D_4 s e^{-s\tau_1} + D_5 s e^{-s\tau_2} + D_6 e^{-s\tau_1} + D_7 e^{-s\tau_2} + D_8 e^{-s(\tau_1+\tau_2)} \tag{2.3.17}$$

and

$$\begin{aligned} D_1 &= aX^* + h + \beta, \\ D_2 &= a\beta X^* + h\beta + \beta X^* - (X^*)^2, \\ D_3 &= a\beta(X^*)^2 + \beta h X^* - a(X^*)^3 - h(X^*)^2, \\ D_4 &= X^* Y^*, \\ D_5 &= -a(d+Y^*)X^* - aX^*Y^*, \\ D_6 &= X^*Y^*(aX^* + h), \\ D_7 &= -X^*[a(d+Y^*)X^* - aX^*Y^*], \\ D_8 &= X^*[-aZ^* + a(d+Y^*)Y^*] \end{aligned}$$

and η_0 is the smallest positive root of Eq. (2.3.16).

Nyquist criterion (see Freedman and Rao 1983; Liao 2005) implies that if η_0 is a solution of Eq. (2.3.16), then

$$\begin{aligned} &-D_1\eta_0^2 + D_3 + D_4\eta_0 \sin(\eta_0\tau_1) + D_5\eta_0 \sin(\eta_0\tau_2) + D_6 \cos(\eta_0\tau_1) \\ &+D_7 \cos(\eta_0\tau_2) + D_8 \cos\{\eta_0(\tau_1+\tau_2)\} = 0 \end{aligned} \tag{2.3.18}$$

Next, we wish to find an upper bound for η (say, η_+) independent of τ_1, τ_2 such that (2.3.15) and (2.3.16) hold $\forall \eta$, $0 \le \eta \le \eta_+$ and hence in particular for η_0 when $\eta_0 \le \eta_+$.

Maximizing (2.3.18) with $|\sin(\eta_0\tau_1)| \le 1$, $|\sin(\eta_0\tau_2)| \le 1$, $|\cos(\eta_0\tau_1)| \le 1$, $|\cos(\eta_0\tau_2)| \le 1$, $|\cos\{\eta_0(\tau_1+\tau_2)\}| \le 1$, we get,

$$D_1\eta_0^2 - (|D_4| + |D_5|)\eta_0 - (|D_3| + |D_6| + |D_7| + |D_8|) = 0 \tag{2.3.19}$$

If η_+ is a positive root of (2.3.19), then we obtain

$$\eta_+ = \frac{1}{2D_1}[(|D_4| + |D_5|) + \{(|D_4| + |D_5|)^2 + 4D_1(|D_3| + |D_6| + |D_7| + |D_8|)\}^{\frac{1}{2}}] \tag{2.3.20}$$

Then clearly, $\eta_+ \ge \eta_0$.

Equation (2.3.15) indicates that the following inequality should hold:

$$\varphi(\tau_1, \tau_2, \eta_0) > \psi(\tau_1, \tau_2, \eta_0),$$

where

$$\varphi(\tau_1, \tau_2, \eta_0) = -\eta_0[\eta_0^2 - D_2 - D_4\cos(\eta_0\tau_1) - D_5\cos(\eta_0\tau_2)] \tag{2.3.21}$$

and

$$\psi(\tau_1, \tau_2, \eta_0) = D_6\sin(\eta_0\tau_1) + D_7\sin(\eta_0\tau_2) + D_8\sin\{\eta_0(\tau_1+\tau_2)\}. \tag{2.3.22}$$

If we can find $\tilde{\varphi}(\tau_1, \tau_2) > \tilde{\psi}(\tau_1, \tau_2)$ such that

$$\frac{\varphi(\tau_1, \tau_2, \eta_0)}{(\tau_1+\tau_2)\eta_0} \geq \tilde{\varphi}(\tau_1, \tau_2) > \tilde{\psi}(\tau_1, \tau_2) \geq \frac{\psi(\tau_1, \tau_2, \eta_0)}{(\tau_1+\tau_2)\eta_0}, \tag{2.3.23}$$

where $0 < \eta_0 < \eta_+$, then the Nyquist criterion holds. By Eq. (2.3.23) we can estimate the values of τ_1 and τ_2.

$$\frac{\psi(\tau_1, \tau_2, \eta_0)}{(\tau_1+\tau_2)\eta_0} = \frac{D_6\sin(\eta_0\tau_1) + D_7\sin(\eta_0\tau_2) + D_8\sin\{\eta_0(\tau_1+\tau_2)\}}{(\tau_1+\tau_2)\eta_0}$$
$$\leq \mid D_6 \mid + \mid D_7 \mid + \mid D_8 \mid$$

We choose,

$$\tilde{\psi}(\tau_1, \tau_2) = \mid D_6 \mid + \mid D_7 \mid + \mid D_8 \mid . \tag{2.3.24}$$

Now,

$$\frac{\varphi(\tau_1, \tau_2, \eta_0)}{(\tau_1+\tau_2)\eta_0} = \frac{\eta_0\{D_2 + D_4\cos(\eta_0\tau_1) + D_5\cos(\eta_0\tau_2) - \eta_0^2\}}{(\tau_1+\tau_2)\eta_0}$$
$$\geq \frac{\{D_2 + D_4\cos(\eta_+\tau_1) + \mid D_5 \mid - \eta_+^2\}}{(\tau_1+\tau_2)}$$

We can choose,

$$\tilde{\varphi}(\tau_1, \tau_2) = \frac{\{D_2 + D_4\cos(\eta_+\tau_1) + \mid D_5 \mid - \eta_+^2\}}{(\tau_1+\tau_2)}. \tag{2.3.25}$$

From (2.3.23) we get,

$$\tilde{\varphi}(\tau_1, \tau_2) = \frac{\{D_2 + D_4\cos(\eta_+\tau_1) + \mid D_5 \mid - \eta_+^2\}}{(\tau_1+\tau_2)} > \mid D_6 \mid + \mid D_7 \mid + \mid D_8 \mid = \tilde{\psi}(\tau_1, \tau_2).$$

Thus

$$\begin{aligned}\tau_1+\tau_2 &< \frac{D_2+D_4\cos(\eta_+\tau_1)+\mid D_5\mid-\eta_+^2}{\mid D_6\mid+\mid D_7\mid+\mid D_8\mid}\\ &\leq \frac{\mid D_2\mid+\mid D_4\mid+\mid D_5\mid-\eta_+^2}{\mid D_6\mid+\mid D_7\mid+\mid D_8\mid}\end{aligned} \tag{2.3.26}$$

So, we come to the following theorem:

Theorem 2.9 *The delayed model (2.3.12a, 2.3.12b) will be locally asymptotically stable at $P^*(X^*, Y^*, Z^*)$ together with the conditions (2.3.9) if*

$$\mid D_2\mid+\mid D_4\mid+\mid D_5\mid > \eta_+^2$$

and

$$0<\tau_1+\tau_2<\frac{\mid D_2\mid+\mid D_4\mid+\mid D_5\mid-\eta_+^2}{\mid D_6\mid+\mid D_7\mid+\mid D_8\mid}.$$

2.3.6 *Analysis of Existence of Hopf Bifurcation*

Now, we will study the existence of Hopf bifurcation of system (2.3.12a, 2.3.12b) by choosing one of the delays as a bifurcation parameter, say, τ_2 as a bifurcation parameter. The characteristic equation corresponding to system (2.3.12a, 2.3.12b) is

$$\begin{aligned}&\lambda^3+D_1\lambda^2+D_2\lambda+D_3+D_4\lambda e^{-\lambda\tau_1}+D_5\lambda e^{-\lambda\tau_2}+D_6e^{-\lambda\tau_1}\\ &+D_7e^{-\lambda\tau_2}+D_8e^{-\lambda(\tau_1+\tau_2)}=0\end{aligned} \tag{2.3.27}$$

Hence, $\lambda=\pm i\eta$ is a simple root of (2.3.27) at $\tau_2=\tau_2^c$. By separating real and imaginary parts, we obtain

$$\begin{aligned}&-D_1\eta^2+D_3+D_4\eta\sin(\eta\tau_1)+D_5\eta\sin(\eta\tau_2)\\ &+D_6\cos(\eta\tau_1)+D_7\cos(\eta\tau_2)+D_8\cos\{\eta(\tau_1+\tau_2)\}=0,\end{aligned} \tag{2.3.28}$$

and

$$\begin{aligned}&-\eta^3+D_2\eta+D_4\eta\cos(\eta\tau_1)+D_5\eta\cos(\eta\tau_2)\\ &-D_6\sin(\eta\tau_1)-D_7\sin(\eta\tau_2)+D_8\sin\{\eta(\tau_1+\tau_2)\}=0.\end{aligned} \tag{2.3.29}$$

Differentiating Eq. (2.3.27) implicitly with respect to τ_2, we obtain

$$\begin{aligned}\left(\frac{d\lambda}{d\tau_2}\right)^{-1} &= \frac{D_5}{\lambda(D_5\lambda+D_7+D_8e^{-\lambda\tau_1})}+\frac{D_8\tau_1e^{-\lambda\tau_1}}{\lambda(D_5\lambda+D_7+D_8e^{-\lambda\tau_1})}\\ &+\frac{3\lambda^2+2D_1\lambda+D_2}{\lambda e^{-\lambda\tau_2}(D_5\lambda+D_7+D_8e^{-\lambda\tau_1})}+\frac{e^{-\lambda\tau_1}(D_4-D_4\lambda\tau_1-D_6\tau_1)}{\lambda e^{-\lambda\tau_2}(D_5\lambda+D_7+D_8e^{-\lambda\tau_1})}-\frac{\tau_2}{\lambda}\end{aligned} \tag{2.3.30}$$

Now,

$$\text{Re}\left(\frac{d\lambda}{d\tau_2}\right)^{-1}_{\lambda=i\eta_0} = \frac{1}{\Theta_1^2+\Theta_2^2}(\Theta_1\Theta_3+\Theta_2\Theta_4),$$

where,

$$\begin{aligned}
\Theta_1 &= -D_5\eta_0^2 + D_8\eta_0 \sin(\eta_0\tau_1),\\
\Theta_2 &= D_7\eta_0 + D_8\eta_0\cos(\eta_0\tau_1),\\
\Theta_3 &= D_5 - D_8\tau_1\cos(\eta_0\tau_1) + (-3\eta_0^2 + D_2)\cos(\eta_0\tau_2) - 2D_1\eta_0\sin(\eta_0\tau_2)\\
&\quad +\{(D_4 - D_6\tau_1)\cos(\eta_0\tau_1) - \eta_0\tau_1 D_4\sin(\eta_0\tau_1)\}\cos(\eta_0\tau_2)\\
&\quad +\{(D_4 - D_6\tau_1)\sin(\eta_0\tau_1) + \eta_0\tau_1 D_4\cos(\eta_0\tau_1)\}\sin(\eta_0\tau_2)\\
\Theta_4 &= D_8\tau_1\sin(\eta_0\tau_1) + (-3\eta_0^2 + D_2)\sin(\eta_0\tau_2) + 2D_1\eta_0\cos(\eta_0\tau_2)\\
&\quad +\{(D_4 - D_6\tau_1)\cos(\eta_0\tau_1) - \eta_0\tau_1 D_4\sin(\eta_0\tau_1)\}\sin(\eta_0\tau_2)\\
&\quad -\{(D_4 - D_6\tau_1)\sin(\eta_0\tau_1) + \eta_0\tau_1 D_4\sin(\eta_0\tau_1)\}\cos(\eta_0\tau_2)
\end{aligned}$$

Clearly, the sign of $\Theta_1^2+\Theta_2^2$ in (2.3.30) is positive. Because of the coherence of the sign of $\Theta_1\Theta_3+\Theta_2\Theta_4$ and $\text{Re}\left(\frac{d\lambda}{d\tau_2}\right)^{-1}_{\lambda=i\eta_0}$, we get, if $\Theta_1\Theta_3+\Theta_2\Theta_4 > 0$, then $\text{Re}\left(\frac{d\lambda}{d\tau_2}\right)^{-1}_{\lambda=i\eta_0} > 0$.

Therefore, transversality condition holds and Hopf bifurcation occurs at $\tau_2 = \tau_2^c$, provided $(\Theta_1\Theta_3+\Theta_2\Theta_4) > 0$.

Summarizing the above analysis, we get the following theorem:

Theorem 2.10 *If there exists $\tau_2 = \tau_2^c$ such that (2.3.28) and (2.3.29) hold and $Re\left(\frac{d\lambda}{d\tau_2}\right)^{-1}_{\lambda=i\eta_0} > 0$, that is, $(\Theta_1\Theta_3+\Theta_2\Theta_4) > 0$, then a Hopf bifurcation occurs at $P^*(X^*, Y^*, Z^*)$ as τ_2 passes through τ_2^c.*

2.3.7 Numerical Simulations

In this section, we present computer simulation of some important analytical results discussed in the previous sections. We take the parameters of the system as given in Table 2.1. Using those values under zero exogenous input, i.e. $u = 0$, it is observed that the equilibrium $P_0(0, 0, 0)$ becomes unstable and interior equilibrium $P^*(X^*, Y^*, Z^*)$ is not feasible. Only the axial equilibrium $P_1(\alpha, 0, 0) = (2.28, 0, 0)$ is locally asymptotically stable, which is shown in Fig. 2.1a. Therefore, Theorem 2.3 is verified. When $u = 1.25$ (i.e. nonzero exogenous input) using the parameter values given in Table 2.2, it is observed that the equilibrium $P_2 = (0, 4.28082, 12.5)$ exists and locally asymptotically stable with $h\alpha\beta < u$, but the interior equilibrium P^* does not exist, which is shown in Fig. 2.1b. On the other hand, when $u = 1.25$ using the parameter values given in Table 2.1, it is found that $h\alpha\beta > u, a(\beta - \alpha - d) > h$, and $P_2 = (0, 2.14041, 6.25)$ becomes unstable and

Table 2.1 Parametric values for Fig. 2.1 (a)

Parameter	Values
α	2.28
β	2.92
a	4.14
d	0.54
h	0.2

Table 2.2 Parametric values for Fig. 2.1 (b)

Parameter	Values
α	2.28
β	2.92
a	4.14
d	0.54
h	0.1

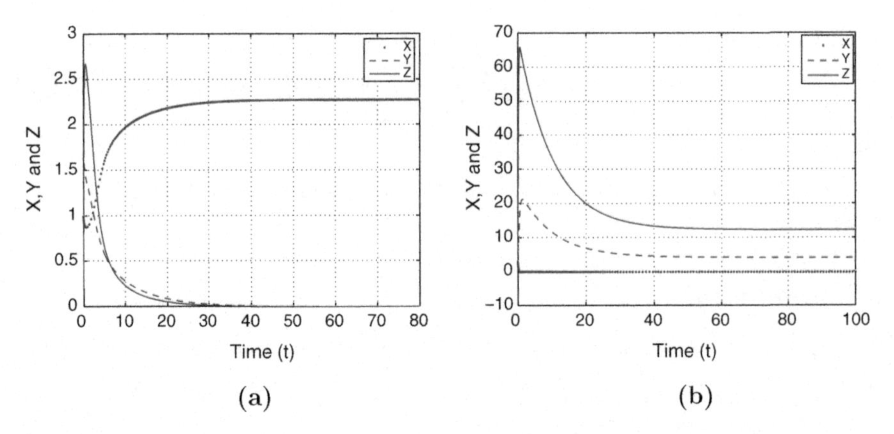

Fig. 2.1 **a** Time series plot of X, Y, Z for zero exogenous input ($u = 0$) with $X(0) = 1$, $Y(0) = 2$, $Z(0) = 0.2$, parameter values are given in Table 2.1. **b** Time series plot of X, Y, Z for nonzero exogenous input ($u = 1.25$) and $h\alpha\beta < u$ with $X(0) = 3$, $Y(0) = 5$, $Z(0) = 8$, parameter values are given in Table 2.2

$P^*(X^*, Y^*, Z^*) = (0.43227, 1.84773, 4.59665)$ exists, which is shown in Fig. 2.2a. Therefore, Theorem 2.4 is verified.

Using the parameter values given in Table 2.1 with $u = 1.25$, it is observed that the conditions of Theorem 2.5 are satisfied as $\Upsilon_1 = 4.9096 > 0$, $\Upsilon_3 = 0.121507 > 0$, $\Upsilon_1\Upsilon_2 - \Upsilon_3 = 0.388821 > 0$. Therefore, by Theorem 2.5, P^* is locally asymptotically stable. Figure 2.2a shows that X, Y and Z approach to their steady-state values X^*, Y^* and Z^*, respectively, in finite time. The XY-plane, YZ-plane, XZ-plane and XYZ-plane projections of the solution are shown in Fig. 2.2b–e, respectively. Here, the conditions of Theorem 2.6 are also satisfied and consequently P^* is globally asymptotically stable. The phase diagram is shown in Fig. 2.2f.

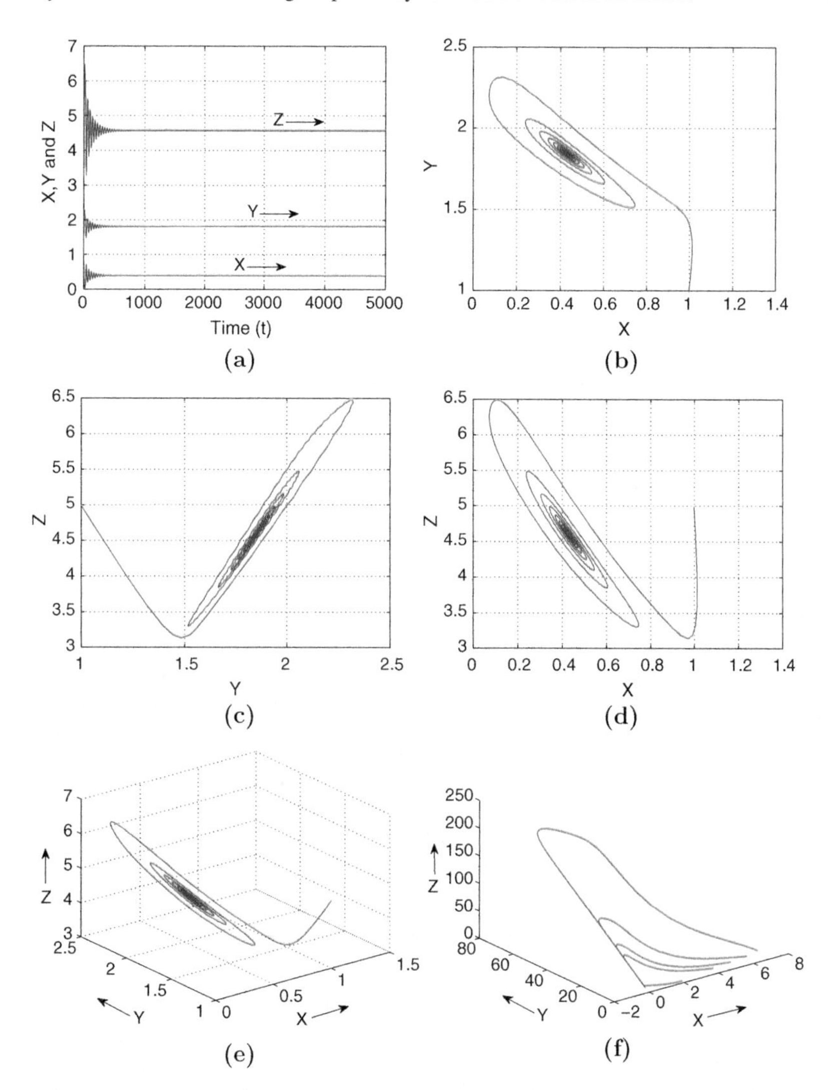

Fig. 2.2 **a** Time series plot of X, Y, Z for nonzero exogenous input ($u = 1.25$) and $h\alpha\beta > u$ with $X(0) = 1$, $Y(0) = 1$, $Z(0) = 5$, parameter values are given in Table 2.1. **b** XY-plane projection of the solution for nonzero exogenous input ($u = 1.25$) with $X(0) = 1$, $Y(0) = 1$, $Z(0) = 5$, parameter values are given in Table 2.1. **c** YZ-plane projection of the solution for nonzero exogenous input ($u = 1.25$) with $X(0) = 1$, $Y(0) = 1$, $Z(0) = 5$, parameter values are given in Table 2.1. **d** XZ-plane projection of the solution for nonzero exogenous input ($u = 1.25$) with $X(0) = 1$, $Y(0) = 1$, $Z(0) = 5$, parameter values are given in Table 2.1. **e** XYZ-plane projection of the solution for nonzero exogenous input ($u = 1.25$) with $X(0) = 1$, $Y(0) = 1$, $Z(0) = 5$, parameter values are given in Table 2.1. **f** Global stability of $P^*(X^*, Y^*, Z^*)$ for nonzero exogenous input ($u = 1.25$) with $(X(0), Y(0), Z(0)) = (2, 1, 3), (4, 3, 5), (6, 4, 8), (5, 3, 8), (7, 8, 4)$, parameter values are given in Table 2.1

2.3.8 Discussion

In this article, we have developed a single-species population model in the polluted environment by means of ordinary differential equations in terms of their concentrations. Boundedness, positivity, stability analysis of the model (2.3.3) with (2.3.4) at various equilibrium points are discussed thoroughly, which implies that the system is ecologically well-behaved. The analysis of the existence of various equilibrium points under zero exogenous toxicant input and nonzero exogenous toxicant input leads us to Theorems 2.3 and 2.4, respectively. Theorem 2.3 states that under zero exogenous input the trivial equilibrium P_0 is unstable, the interior equilibrium P^* is not feasible and only the axial equilibrium P_1 is locally asymptotically stable. On the other hand, when exogenous input is nonzero, the existence of the interior equilibrium P^* depends on some conditions and the equilibrium P_2 is locally asymptotically stable under some conditions. The stability criteria for P^* given in Theorems 2.5 and 2.6 are the conditions, respectively, for locally and globally stable coexistence of the population biomass, population toxicant and environmental toxicant.

It is proved by several researchers that the effect of time delays must be taken into account in the population models to make them ecologically more meaningful and useful. It seems also reasonable that the effect of the environmental toxicant on the organismal activities and the population growth will not be instantaneous. It must be mediated with some discrete time lag required for incubation. From this point of view, we have formulated double discrete time delayed model given in (2.3.12a, 2.3.12b), where the delays may be looked upon as the reaction time of the environment toxicant on the population biomass and the population toxicant.

The analysis of the delayed model (2.3.12a, 2.3.12b) shows more complicated behaviour than their non-delayed counterpart. The stability of the double-delayed model (2.3.12a, 2.3.12b) is investigated by the Nyquist criteria which leads us to Theorem 2.7. By choosing one of the delays as a bifurcation parameter, the model (2.3.12a, 2.3.12b) is found to undergo a Hopf bifurcation under some conditions stated in Theorem 2.8. Our important mathematical findings are numerically verified by using MATLAB, which show the mathematical and ecological reliability of the proposed model.

References

Birkhoff, G., Rota, G.C.: Ordinary Differential Equations. Ginn, Boston (1982)

Buonomo, B., Liddo, A.D., Sgura, I.: A diffusive-connective model for the dynamics of population-toxicant intentions: some analytical and numerical results. Math. Biosci. **157**, 37–64 (1999)

Erbe, L.H., Rao, V.S.H., Freedman, H.: Three species food chain models with mutual interference and time delays. Math. Biosci. **80**, 57–80 (1986)

Freedman, H.I.: Deterministic Mathematical Models in Population Ecology. HIFR Consulting Ltd., Edmonton (1987)

Freedman, H., Rao, V.S.H.: The trade-off between mutual interference and time lags in predator-prey systems. Bull. Math. Biol. **45**, 991–1004 (1983)

Hale, J.K.: Theory of Functional Differential Equations. Springer-Verlag, New York (1977)
Liao, X.: Hopf and resonant codimension two bifurcation in Van Dar Pol equation with two time delays. Chaos Solitons Fractals **23**, 857–871 (2005)
Liu, W.: Criterion of Hopf bifurcations without using eigen values. J. Math. Anal. Appl. **182**, 250–256 (1994)
May, R.M.: Stability in randomly fluctuating versus deterministic environments. Am. Nat. **107**, 621–650 (1973)
White, B.S.: The effects of a rapidly fluctuating random environment on systems of interacting species. SIAM J. Appl. Math. **32**, 666–693 (1977)

Chapter 3
Analysis of Nonautonomous Two Species Systems in a Polluted Environment

3.1 Introduction

In this chapter, our matter of discussion is the dynamical behaviour of different nonautonomous two species systems in a polluted environment. It is observed that a large quantity of toxicant and contaminants enter into ecosystems one after another due to the rapid development of modern industries and agricultures. Of late, many countries have already realized that the pollution of the environment is a very serious problem and urgent steps should be needed. One of the most important and meaningful questions in mathematical ecology is the uniformly persistent (the survival of all species which exist initially) and extinction of a population in an environment that is polluted. Organisms are mostly exposed to a polluted environment and taking up toxicant. The change in environment caused by pollution affects the long-term survival of most biological systems, humans and biodiversity of the habitat. The question of the adverse effects of pollutants and toxicants on ecological communities is of tremendous important from both an environmental and conservational point of view. Allowing the parameters of the systems to be time dependent (*nonautonomous*) is, theoretically, a way of modelling environmental variations, changes in growth rates and in general effects of perturbation of the underlying system. We have derived sufficient conditions on the permanence of the systems. Global asymptotic behaviour of these systems by establishing suitable bounded functions has been investigated.

Nonautonomous phenomenon often arises in many realistic ecoepidemiological models. It happens mainly due to the seasonal variety, which makes the population to behave periodically. Pathogen contact rate and infectivity differ seasonally and is generally larger in spring and autumn than in summer and winter. This periodic pattern is related to moderate temperatures in spring and autumn which improves pathogen survival and favours high insect activity. As biological and environmental parameters are naturally subject to fluctuation in time, the effects of a periodically varying environment are considered as important selective forces on systems in a fluctuating environment. For investigating this kind of phenomenon, in the model, the

© Springer Nature Singapore Pte Ltd. 2021

G. Samanta, *Deterministic, Stochastic and Thermodynamic Modelling of some Interacting Species*, Forum for Interdisciplinary Mathematics,
https://doi.org/10.1007/978-981-16-6312-3_3

coefficients should be periodic functions, then the system is called periodic system. The nonautonomous two species systems can be considered as an extension of the periodic two species systems. So, the research on the nonautonomous two species dynamical systems is also very important.

3.2 Two Species Systems in a Polluted Environment

Since species do not exist alone in nature, the study of the persistence–extinction threshold of each population in systems of two or more interacting species subject to organismal and environmental toxicants is very ecologically significant. In this article, we have considered a two-species nonautonomous Lotka–Volterra model with organismal and environmental toxicant effects. The investigation of the global asymptotic behaviour is based on finding bounded functions $H_i > 0$ $(i = 1, 2)$, satisfying equations which are suitable perturbations of original equations. We have proved that each population for competition, predation and cooperation systems, respectively, are uniformly persistent (permanent) under stated appropriate conditions. We have derived sufficient conditions by constructing an appropriate function (called Lyapunov function) to confirm that if each of competition, predation and cooperation systems, respectively, admits a positive periodic solution, then it is globally asymptotically stable.

3.2.1 The Basic Mathematical Model

This study is based on the hypothesis of a total spatial homogeneous environment and no migration. The following model is considered to describe the dynamics of population of two species in a polluted environment where the state variables $X_1(t)$ and $X_2(t)$ represent the densities of the first and second species at time t, respectively, $\xi_0(t)$ and $\xi_e(t)$ stand for the concentrations of toxicant in the organism and in the environment at time t, respectively. It is assumed that the individuals in the two species have the identical organismal toxicant concentration at time t.

$$
\begin{aligned}
\frac{dX_1(t)}{dt} &= X_1(t)[b_{10}(t) - r_1(t)\xi_0(t) - b_{11}(t)X_1(t) - b_{12}(t)X_2(t)], \\
\frac{dX_2(t)}{dt} &= X_2(t)[b_{20}(t) - r_2(t)\xi_0(t) - b_{21}(t)X_1(t) - b_{22}(t)X_2(t)], \\
\frac{d\xi_0(t)}{dt} &= k(t)\xi_e(t) + \rho(t) - g(t)\xi_0(t) - \delta(t)\xi_0(t), \\
\frac{d\xi_e(t)}{dt} &= -\psi(t)\xi_e(t) + \eta(t)
\end{aligned}
\tag{3.2.1}
$$

The initial conditions are:

$$X_{10} = X_1(0) > 0,\ X_{20} = X_2(0) > 0$$
$$0 \le \xi_0(0) \le 1,\ 0 \le \xi_e(0) \le 1 \tag{3.2.2}$$

Here, $b_{i0}(t)$, $r_i(t)$ and $b_{ij}(t)$ $(i, j = 1, 2)$ are continuous and bounded functions on $[0, \infty)$, where $b_{i0}(t)$ and $r_i(t)$ represent, respectively, the intrinsic growth rate and response to the organismal toxicant concentration of species i, and $b_{ij}(t)$ measures the action of species j upon the growth rate of species i (in particular, and $b_{ii}(t)$ represents the intraspecific competition coefficient of species i). It is assumed that the intrinsic growth rates of the population are decreased linearly due to the effect of toxicant. The organismal net uptake of toxicant from the environment $(k(t)\xi_e(t))$ and the food chain $(\rho(t))$ are denoted in the first two terms on the right of the third equation in system (3.2.1), respectively, the organismal net loss of toxicant due to metabolic processing and other causes are represented by the third and fourth terms. Environmental toxicant uptake rate per unit mass organism is denoted by $k(t)$, uptake rate of toxicant in food per unit mass organism is denoted by $\rho(t)$, and the egestion and depuration rates of the toxicant in the organism are represented by $-g(t)\xi_0(t)$ and $-\delta(t)\xi_0(t)$, respectively. The first term $(-\psi(t)\xi_e(t))$ on the right of the fourth equation in system (3.2.1) represents the toxicant loss from the environment itself by biological transformation, chemical hydrolysis, volatilization, microbial degradation, photosynthetic degradation and so on, and $\eta(t)$ represents the exogenous rate of input of toxicant into the environment. The change of toxicants in the environment that comes from uptake and egestion by the organisms can be neglected due to the huge capacity of the environment. The coefficients $k(t)$, $\rho(t)$, $g(t)$, $\delta(t)$, $\psi(t)$ and $\eta(t)$ are assumed to be nonnegative, continuous and bounded functions on $[0, \infty)$.

Let $\phi^l = \inf_{t\ge0}\phi(t)$, $\phi^u = \sup_{t\ge0}\phi(t)$, for a continuous and bounded function $\phi(t)$ defined in the interval $[0, +\infty)$.

Each of $\xi_0(t)$ and $\xi_e(t)$ is a concentration, and so these variables cannot be greater than 1; that is, some conditions should be needed such that $0 \le \xi_0(t) \le 1$, $0 \le \xi_e(t) \le 1$, for all $t \ge 0$.

Theorem 3.1 *Every solution of system (3.2.1) with initial conditions (3.2.2) uniquely exists and* $X_1(t) > 0$, $X_2(t) > 0$, $\xi_0(t) \ge 0$, $\xi_e(t) \ge 0$, *for all* $t \in [0, \infty)$.

Proof Since the right-hand side of system (3.2.1) is completely continuous and locally Lipschitzian on $\mathbb{C}$, the solution $(X_1(t), X_2(t), \xi_0(t), \xi_e(t))$ of (3.2.1) with initial conditions (3.2.2) exists and is unique on $[0, \beta)$, where $0 < \beta \le +\infty$ (Hale 1977; Chap. 2). From (3.2.1) and (3.2.2):

$$X_1(t) = X_1(0)\exp\left\{\int_0^t [b_{10}(s) - r_1(s)\xi_0(s) - b_{11}(s)X_1(s) - b_{12}(s)X_2(s)]\mathrm{d}s\right\} > 0,$$
$$\forall\, t \ge 0.$$

$$X_2(t) = X_2(0)\exp\left\{\int_0^t [b_{20}(s) - r_2(s)\xi_0(s) - b_{21}(s)X_1(s) - b_{22}(s)X_2(s)]\mathrm{d}s\right\} > 0,$$
$$\forall\, t \geq 0.$$

$$\xi_0(t) = \mathrm{e}^{-\int_0^t \{\delta(s)+g(s)\}\mathrm{d}s}\left\{\int_0^t \{k(s)\xi_\mathrm{e}(s) + \rho(s)\}\mathrm{e}^{\int_0^s \{\delta(s)+g(s)\}\mathrm{d}s}\mathrm{d}s + \xi_0(0)\right\} \geq 0,$$
$$\forall\, t \geq 0.$$

$$\xi_\mathrm{e}(t) = \mathrm{e}^{-\int_0^t \psi(s)\mathrm{d}s}\left\{\int_0^t \eta(s)\mathrm{e}^{\int_0^s \psi(s)\mathrm{d}s}\mathrm{d}s + \xi_\mathrm{e}(0)\right\} \geq 0,\ \forall\, t \geq 0.$$

□

Theorem 3.2 *From (3.2.1) with (3.2.2): if* $k^u + \rho^u \leq g^l + \delta^l$, $\eta^u \leq \psi^l$, *then* $0 \leq \xi_0(t) \leq 1, 0 \leq \xi_e(t) \leq 1$, *for all* $t \geq 0$.

Proof According to Theorem 3.1, we have $\xi_0(t) \geq 0$, $\xi_\mathrm{e}(t) \geq 0$, for all $t \geq 0$. Now, we have to prove that $\xi_0(t) \leq 1$, $\xi_\mathrm{e}(t) \leq 1$, for all $t \in [0, \infty)$.

If the conclusion is false, then the maximum interval is $[0, T]$ such that $0 \leq \xi_0(t) \leq 1, 0 \leq \xi_\mathrm{e}(t) \leq 1, \forall\, t \in [0, T]$ and at least one of the following situations will arise:

$$\text{(i) } \xi_0(T) = 1, \xi_\mathrm{e}(T) < 1;$$

$$\text{(ii) } \xi_0(T) < 1, \xi_\mathrm{e}(T) = 1;$$

$$\text{(iii) } \xi_0(T) = \xi_\mathrm{e}(T) = 1.$$

Let us prove that none of these situations is true.

(i) $\xi_0(T) = 1, \xi_\mathrm{e}(T) < 1$; using the condition $k^u + \rho^u \leq g^l + \delta^l$, we have

$$\begin{aligned}\frac{\mathrm{d}\xi_0(t)}{\mathrm{d}t}\Big|_{t=T} &= k(T)\xi_\mathrm{e}(T) + \rho(T) - g(T)\xi_0(T) - \delta(T)\xi_0(T)\\ &< k(T) + \rho(T) - g(T) - \delta(T)\\ &\leq k^u + \rho^u - g^l - \delta^l \leq 0,\end{aligned}$$

thus $\exists\, t_1 > 0$, s.t. $0 \leq \xi_0(t) \leq 1,\ 0 \leq \xi_\mathrm{e}(t) \leq 1, \forall\, t \in [T, T + t_1]$. This contradicts the definition of the interval $[0, T]$, and so there is no T such that $0 \leq \xi_0(t) \leq 1, 0 \leq \xi_\mathrm{e}(t) \leq 1, \forall\, t \in [0, T]$ with $\xi_0(T) = 1, \xi_\mathrm{e}(T) < 1$.

With the same reasoning as in (i), for situations (ii) and (iii), as far as t which keeps $0 \leq \xi_0(t) \leq 1$ and $0 \leq \xi_\mathrm{e}(t) \leq 1$ is concerned, the interval $[0, T]$ can be extended rightwards. This contradicts the property of T, hence $0 \leq \xi_0(t) \leq 1,\ 0 \leq \xi_\mathrm{e}(t) \leq 1, \forall\, t \geq 0$. Thus, the proof is completed. □

Putting the expression of $\xi_e(t)$ in $\xi_0(t)$, we can express $\xi_0(t)$ in term of some bounded continuous functions. Therefore, the system (3.2.1) with (3.2.2) may be simplified as follows:

$$\begin{aligned}\frac{dX_1(t)}{dt} &= X_1(t)[b_1(t) - b_{11}(t)X_1(t) - b_{12}(t)X_2(t)],\\ \frac{dX_2(t)}{dt} &= X_2(t)[b_2(t) - b_{21}(t)X_1(t) - b_{22}(t)X_2(t)],\end{aligned} \tag{3.2.3}$$

where

$$b_1(t) = b_{10}(t) - r_1(t)\xi_0(t),\ b_2(t) = b_{20}(t) - r_2(t)\xi_0(t),$$

with initial conditions:

$$X_{10} = X_1(0) > 0,\ X_{20} = X_2(0) > 0. \tag{3.2.4}$$

3.2.2 Global Stability of System (3.2.3)

Theorem 3.3 *Assume that there exist $H_i(t) > 0, i = 1, 2,$ continuously differentiable for all $t \in [0, \infty)$ such that*

$$\frac{dH_i(t)}{dt} = H_i(t)\left[b_i(t) - \sum_{j=1}^{2} b_{ij}(t)H_j(t)\right] + H_i(t)h_i(t), \tag{3.2.5}$$

where $H_i(t)$ and $h_i(t)$ are bounded on $[0, \infty)$ and $\int_0^{\infty} \mid h_i(t) \mid dt < \infty$, then system (3.2.3) and (3.2.4) is globally asymptotically stable.

Proof Consider the function $\Omega(t) = \Omega(t, X_1(t), X_2(t))$ defined by

$$\Omega(t) = \sum_{i=1}^{2} \Omega_i(t) \text{ for } t \geq 0, \text{ where } \Omega_i(t) = \left| \int_{H_i(t)}^{X_i(t)} \frac{ds}{s} \right| \quad (i = 1, 2) \tag{3.2.6}$$

in which $X(t) = (X_1(t), X_2(t))^T$ is a solution of (3.2.3) with (3.2.4) and $H_i(t)$ is a function such that (3.2.5) holds.

Calculating the upper-right derivative $D^+\Omega(t)$ of $\Omega(t)$ along the solution of (3.2.3) with (3.2.4), we get

$$\begin{aligned}
D^{+}\Omega(t) &= \sum_{i=1}^{2} D^{+}\Omega_i(t) \\
&= \sum_{i=1}^{2}\left(\frac{\dot{X}_i(t)}{X_i(t)} - \frac{\dot{H}_i(t)}{H_i(t)}\right)\operatorname{sgn}(X_i(t) - H_i(t)) \\
&= -\sum_{i=1}^{2}\sum_{j=1}^{2} b_{ij}(t) \mid X_j(t) - H_j(t) \mid \\
&\quad - \sum_{i=1}^{2} h_i(t)\operatorname{sgn}(X_i(t) - H_i(t)), \ \text{for } t \geq 0
\end{aligned}$$

$$\Rightarrow D^{+}\Omega(t) \leq -\sum_{i=1}^{2}\sum_{j=1}^{2} b_{ij}(t) \mid X_j(t) - H_j(t) \mid + \sum_{i=1}^{2} \mid h_i(t) \mid, \ \text{for } t \geq 0 \quad (3.2.7)$$

Integrating both sides of (3.2.7) with respect to t, we have

$$\begin{aligned}
\Omega(t) + \sum_{i=1}^{2}\sum_{j=1}^{2}\int_{0}^{t} b_{ij}(\theta) \mid X_j(\theta) - H_j(\theta) \mid d\theta &\leq \Omega(0) \\
+ \sum_{i=1}^{2}\int_{0}^{t} \mid h_i(s) \mid ds, \ \text{for } t \geq 0
\end{aligned}$$

$$\begin{aligned}
\Rightarrow \Omega(t) + \sum_{j=1}^{2}\int_{0}^{t}\left(\sum_{i=1}^{2} b_{ij}(\theta)\right) \mid X_j(\theta) - H_j(\theta) \mid d\theta &\leq \Omega(0) \\
+ \sum_{i=1}^{2}\int_{0}^{t} \mid h_i(s) \mid ds, \ \text{for } t \geq 0
\end{aligned} \quad (3.2.8)$$

Since all the coefficients in the system are continuous and bounded functions on $[0, \infty)$ and moreover $\int_0^{\infty} \mid h_i(t) \mid dt < \infty$, there exist constants C_j $(j = 1, 2)$ such that

$$\Omega(t) + \sum_{j=1}^{2} C_j \int_{0}^{t} \mid X_j(\theta) - H_j(\theta) \mid d\theta \leq \zeta, \ \text{for some constant } \zeta > 0. \quad (3.2.9)$$

Therefore, $\Omega(t)$ is bounded on $[0, \infty)$, and hence, $\mid \int_{H_i(t)}^{X_i(t)} \frac{ds}{s} \mid$ $(i = 1, 2)$ is bounded on $[0, \infty)$. Since $H_i(t)$ is bounded function on $[0, \infty)$, it follows that

$$X_i(t) \text{ is bounded on } [0, \infty), \ \text{for } i = 1, 2. \quad (3.2.10)$$

Therefore,

$$| X_i(t) - H_i(t) | \text{ is bounded on } [0, \infty), \text{ for } i = 1, 2. \tag{3.2.11}$$

Also, from (3.2.9),

$$\int_0^\infty | X_i(\theta) - H_i(\theta) | \, d\theta < \infty, \; i = 1, 2. \tag{3.2.12}$$

On the other hand, from the hypotheses, it is evident that $\dot{X}_i(t)$ and $\dot{H}_i(t)$ $(i = 1, 2)$ are bounded on $[0, \infty)$. Therefore,

$$| X_i(t) - H_i(t) | \text{ is uniformly continuous on } [0, \infty), \text{ for } i = 1, 2. \tag{3.2.13}$$

From (3.2.11), (3.2.12), and (3.2.13), we get

$$| X_i(t) - H_i(t) | \to 0 \text{ as } t \to \infty, \; (i = 1, 2). \tag{3.2.14}$$

For any two solutions $X(t) = (X_1(t), X_2(t))^{\mathrm{T}}$ and $Y(t) = (Y_1(t), Y_2(t))^{\mathrm{T}}$ of (3.2.3) with initial conditions (3.2.4), one has

$$0 \leq | X_i(t) - Y_i(t) | \leq | X_i(t) - H_i(t) | + | Y_i(t) - H_i(t) |, \; i = 1, 2$$

$$\Rightarrow | X_i(t) - Y_i(t) | \to 0 \text{ as } t \to \infty, \; i = 1, 2. \tag{3.2.15}$$

Therefore, system (3.2.3) with initial conditions (3.2.4) is globally asymptotically stable. This completes the theorem. □

3.2.3 Competition System

Let us consider the following competition system generated by (3.2.3) and (3.2.4):

$$\begin{aligned} \frac{dX_1(t)}{dt} &= X_1(t)[b_1(t) - b_{11}(t)X_1(t) - b_{12}(t)X_2(t)], \\ \frac{dX_2(t)}{dt} &= X_2(t)[b_2(t) - b_{21}(t)X_1(t) - b_{22}(t)X_2(t)], \end{aligned} \tag{3.2.16}$$

where

$$b_1(t) = b_{10}(t) - r_1(t)\xi_0(t), \; b_2(t) = b_{20}(t) - r_2(t)\xi_0(t),$$

with initial conditions:

$$X_{10} = X_1(0) > 0, \; X_{20} = X_2(0) > 0. \tag{3.2.17}$$

The coefficients $b_i(t), b_{ij}(t)$ $(i, j = 1, 2)$ are assumed to be continuous and bounded on $[0, \infty)$ with $\min_{i,j=1,2}\{b_i^l, b_{ij}^l\} > 0$ and $\max_{i,j=1,2}\{b_i^u, b_{ij}^u\} < \infty$.
In this section, we analyse the permanence and global asymptotic stability of periodic solution of system (3.2.16) with (3.2.17).

Uniformly persistent

Here, we would like to discuss the uniformly persistent of the system (3.2.16) with (3.2.17), which demonstrates how this system will survive forever (i.e. will not be going to extinct in time), under some conditions.

Definition The system (3.2.16) is said to be uniformly persistent, in the ecological context, it implies the survival of all species which exist initially. That is, the long-term survival (will not vanish in time) of all components of the system (3.2.16), if there are positive constants ν_i and σ_i $(i = 1, 2)$ such that:

$$\nu_1 \le \liminf_{t\to\infty} X_1(t) \le \limsup_{t\to\infty} X_1(t) \le \sigma_1,$$

$$\nu_2 \le \liminf_{t\to\infty} X_2(t) \le \limsup_{t\to\infty} X_2(t) \le \sigma_2,$$

hold for any solution $(X_1(t), X_2(t))$ of (3.2.16) with (3.2.17). Here, ν_i and σ_i $(i = 1, 2)$ are independent of the initial conditions (3.2.17).

Theorem 3.4 *Let $X(t) = (X_1(t), X_2(t))^T$ be any solution of system (3.2.16) with (3.2.17). Then $\exists\, T_1 > 0$ such that*

$$X_i(t) \le \Delta \; (i = 1, 2), \; \forall\, t \ge T_1, \; \textit{where } \Delta > \Delta^* = \max\left\{\frac{b_1^u}{b_{11}^l}, \frac{b_2^u}{b_{22}^l}\right\}.$$

Proof Let us define $\Omega(t) = \max\{X_1(t), X_2(t)\}$.
Calculating the upper-right derivative of $\Omega(t)$ along the solution of system (3.2.16) with (3.2.17), we have

$$(1) \text{ if } X_1(t) \ge X_2(t), \text{ then } D^+\Omega(t) = \dot{X}_1(t) \le X_1(t)[b_1(t) - b_{11}(t)X_1(t)]$$

$$\le X_1(t)[b_1^u - b_{11}^l X_1(t)] \tag{3.2.18}$$

$$(2) \text{ if } X_1(t) \le X_2(t), \text{ then } D^+\Omega(t) = \dot{X}_2(t) \le X_2(t)[b_2(t) - b_{22}(t)X_2(t)]$$

$$\le X_2(t)[b_2^u - b_{22}^l X_2(t)] \tag{3.2.19}$$

From (3.2.18) and (3.2.19), we can get the following:

(i) If $\max\{X_1(0), X_2(0)\} \leq \Delta$, then $\max\{X_1(t), X_2(t)\} \leq \Delta,\ \forall\, t \geq 0$.

(ii) If $\max\{X_1(0), X_2(0)\} > \Delta$, and let $-\zeta = \max_{i=1,2}\{\Delta(b_i^u - b_{ii}^l\Delta)\}, \zeta > 0$, we consider the following three possibilities:

$$(a)\ \Omega(0) = X_1(0) > \Delta,\ (X_1(0) > X_2(0));$$

$$(b)\ \Omega(0) = X_2(0) > \Delta,\ (X_1(0) < X_2(0));$$

$$(c)\ \Omega(0) = X_1(0) = X_2(0) > \Delta.$$

If (a) holds, then $\exists$ an $\epsilon > 0$, s.t. if $t \in [0, \epsilon)$, $\Omega(t) = X_1(t) > \Delta$, and we have $D^+\Omega(t) = \dot{X}_1(t) < -\zeta < 0$.
If (b) holds, then $\exists$ an $\epsilon > 0$, s.t. if $t \in [0, \epsilon)$, $\Omega(t) = X_2(t) > \Delta$, and we have $D^+\Omega(t) = \dot{X}_2(t) < -\zeta < 0$.
If (c) holds, then $\exists$ an $\epsilon > 0$, s.t. if $t \in [0, \epsilon)$, $\Omega(t) = X_i(t) > \Delta$, $(i = 1, 2)$. Similar to (a) and (b), we have $D^+\Omega(t) = \dot{X}_i(t) < -\zeta < 0$ $(i = 1, 2)$.
From examining (a), (b) and (c), we conclude that if $\Omega(0) > \Delta$, then $\exists\, T_1 > 0$ s.t.

$$\Omega(t) = \max\{X_1(t), X_2(t)\} \leq \Delta, \forall\, t \geq T_1$$

Hence, the theorem is proved. □

Theorem 3.5 *Suppose that system (3.2.16) with (3.2.17) satisfies the following conditions:*

$$\max\left\{\frac{b_1^u}{b_{11}^l}, \frac{b_2^u}{b_{22}^l}\right\} < \Delta < \min\left\{\frac{b_1^l}{b_{12}^u}, \frac{b_2^l}{b_{21}^u}\right\},$$

then it is uniformly persistent.

Proof Suppose $X(t) = (X_1(t), X_2(t))^{\mathrm{T}}$ is a solution of (3.2.16) with (3.2.17). Therefore,

$$\dot{X}_1(t) \geq X_1(t)[b_1^l - b_{11}^u X_1(t) - b_{12}^u\Delta],\ \forall\, t \geq T_1,$$

$$\dot{X}_2(t) \geq X_2(t)[b_2^l - b_{22}^u X_2(t) - b_{21}^u\Delta],\ \forall\, t \geq T_1.$$

From the given conditions: $b_i^l - b_{i\,3-i}^u\Delta > 0,\ i = 1, 2$.
Let $\sigma^* = \min\left\{\dfrac{r_1^l - b_{12}^u\Delta}{b_{11}^u}, \dfrac{b_2^l - b_{21}^u\Delta}{b_{22}^u}\right\}$, we take σ as $0 < \sigma < \sigma^*$.

Define $\Omega_1(t) = \min\{X_1(t), X_2(t)\}$ and then calculating the lower-right derivative of $\Omega_1(t)$ along the solution of system (3.2.16) with (3.2.17), we obtain:

$$(1)\text{ if } X_1(t) \leq X_2(t),\ \text{then } D_+\Omega_1(t) = \dot{X}_1(t) \geq X_1(t)[b_1^l - b_{11}^u X_1(t) - b_{12}^u\Delta],\ \forall\, t \geq T_1 \quad (3.2.20)$$

$$(2)\text{ if } X_1(t) \geq X_2(t), \text{ then } D_+\Omega_1(t) = \dot{X}_2(t) \geq X_2(t)[b_2^l - b_{22}^u X_2(t) - b_{21}^u \Delta], \ \forall\, t \geq T_1 \tag{3.2.21}$$

From (3.2.20) and (3.2.21), we have the following:
(i) If $\Omega_1(T_1) = \min\{X_1(T_1), X_2(T_1)\} \geq \sigma$, then $\min\{X_1(t), X_2(t)\} \geq \sigma, \ \forall\, t \geq T_1$.

(ii) If $\Omega_1(T_1) = \min\{X_1(T_1), X_2(T_1)\} < \sigma$, and let $\gamma = \min_{i=1,2}\{X_i(T_1)(b_i^l - b_{ii}^u \sigma - b_{i\,3-i}^u \Delta)\}$, we consider the following cases:

$$(a)\ \Omega_1(T_1) = X_1(T_1) < \sigma, \ (X_1(T_1) < X_2(T_1));$$

$$(b)\ \Omega_1(T_1) = X_2(T_1) < \sigma, \ (X_2(T_1) < X_1(T_1));$$

$$(c)\ \Omega_1(T_1) = X_1(T_1) = X_2(T_1) < \sigma.$$

If (a) holds, then $\exists$ an $\epsilon > 0$, s.t. if $t \in [T_1, T_1 + \epsilon)$, $\Omega_1(t) = X_1(t) < \sigma$, and we have $D_+\Omega_1(t) = \dot{X}_1(t) > \gamma > 0$.
If (b) holds, then $\exists$ an $\epsilon > 0$, s.t. if $t \in [T_1, T_1 + \epsilon)$, $\Omega_1(t) = X_2(t) < \sigma$, and we have $D_+\Omega_1(t) = \dot{X}_2(t) > \gamma > 0$.
If (c) holds, then $\exists$ an $\epsilon > 0$, s.t. if $t \in [T_1, T_1 + \epsilon)$, $\Omega_1(t) = X_i(t) < \sigma$, $(i = 1, 2)$.
Similar to (a) and (b), we have $D_+\Omega_1(t) = \dot{X}_i(t) > \gamma > 0$ $(i = 1, 2)$.
From investigating (a), (b) and (c), we conclude that if $\Omega_1(T_1) = \min\{X_1(T_1), X_2(T_1)\} < \sigma$, then
$\exists\, T_2 > T_1 > 0$ s.t.

$$\Omega_1(t) = \min\{X_1(t), X_2(t)\} \geq \sigma, \forall\, t \geq T_2.$$

From the discussions, we conclude that $\exists\, T_2 > T_1 > 0$ s.t. every solution of system (3.2.16) with (3.2.17) eventually enters and remains in the region $B = \{(X_1, X_2) \mid \sigma \leq X_i \leq \Delta, \ i = 1, 2\}$, $\forall\, t \geq T_2$. Hence, the theorem is proved. □

Global asymptotic stability of periodic solution

From our day-to-day experience, it is known that the biological and environmental parameters are subjected to fluctuation in time. The effects of a periodically varying environment have a crucial selective forces on systems in a fluctuating (random) environment. For investigating this kind of natural phenomenon, in the model, the coefficients should be periodic functions of time.

Let that system (3.2.16) is a periodic system with (3.2.17) satisfying all conditions of Theorem 3.5. The sufficient conditions for all solutions of system (3.2.16) with (3.2.17) to converge to a periodic solution are derived in Theorem 3.8.

Theorem 3.6 *(Brouwer (Hirsch 1980)). Assume that the continuous operator f maps the closed and bounded convex set $C \in R^n$ onto itself, then the operator f has at least one fixed point in set C.*

Theorem 3.7 *If the periodic system (3.2.16) with (3.2.17) satisfies the conditions of Theorem 3.5, then there exists at least one strictly positive periodic solution of this system.*

Proof From Theorems 1.5, 1.6 and 2 in Teng and Chen (1999), it is easy to derive at least one strictly positive periodic solution of system (3.2.16) with (3.2.17). □

Theorem 3.8 *The system (3.2.16) with (3.2.17) is globally asymptotically stable under following conditions:*

$$\max\left\{\frac{b_1^u}{b_{11}^l}, \frac{b_2^u}{b_{22}^l}\right\} < \Delta < \min\left\{\frac{b_1^l}{b_{12}^u}, \frac{b_2^l}{b_{21}^u}\right\} \text{ and } b_{ii}^l > b_{3-i\,i}^u \; (i = 1, 2).$$

Proof Let $(\omega_1(t), \omega_2(t))^{\mathrm{T}}$ is a periodic solution of system (3.2.16) with (3.2.17). Suppose that $(X_1(t), X_2(t))^{\mathrm{T}}$ is any positive solution of (3.2.16) with (3.2.17). Let

$$\Omega(t) = \mid \ln X_1(t) - \ln \omega_1(t) \mid + \mid \ln X_2(t) - \ln \omega_2(t) \mid . \tag{3.2.22}$$

Calculating the upper-right derivative of $\Omega(t)$ along the positive solution of system (3.2.16) with (3.2.17):

$$\begin{aligned} D^+\Omega(t) = &\left(\frac{\dot{X}_1(t)}{X_1(t)} - \frac{\dot{\omega}_1(t)}{\omega_1(t)}\right) \operatorname{sgn}(X_1(t) - \omega_1(t)) \\ &+ \left(\frac{\dot{X}_2(t)}{X_2(t)} - \frac{\dot{\omega}_2(t)}{\omega_2(t)}\right) \operatorname{sgn}(X_2(t) - \omega_2(t)). \end{aligned} \tag{3.2.23}$$

After some simplifications and using the stated conditions $b_{ii}^l > b_{3-i\,i}^u$ $(i = 1, 2)$, we have

$$D^+\Omega(t) \leq -(b_{11}^l - b_{21}^u) \mid X_1(t) - \omega_1(t) \mid -(b_{22}^l - b_{12}^u) \mid X_2(t) - \omega_2(t) \mid . \tag{3.2.24}$$

Integrating both sides of (3.2.24) on $[T_2, t]$:

$$\begin{aligned} \Omega(t) + (b_{11}^l - b_{21}^u) &\int_{T_2}^{t} \mid X_1(s) - \omega_1(s) \mid \mathrm{d}s \\ +(b_{22}^l - b_{12}^u) &\int_{T_2}^{t} \mid X_2(s) - \omega_2(s) \mid \mathrm{d}s \leq \Omega(T_2). \end{aligned} \tag{3.2.25}$$

$$\Rightarrow \Omega(t) \text{ is bounded on } [T_2, \infty) \text{ and also } \int_{T_2}^{\infty} \mid X_i(s) - \omega_i(s) \mid \mathrm{d}s < \infty, \;\; (i = 1, 2) \tag{3.2.26}$$

By Theorem (3.5), $\mid X_i(s) - \omega_i(s) \mid$ $(i = 1, 2)$ are bounded on $[T_2, \infty)$. On the other hand, it is observed that $\dot{\omega}_i(t)$ and $\dot{X}_i(t)$, $(i = 1, 2)$ are bounded for $t \geq T_2$. Therefore, $\mid X_i(s) - \omega_i(s) \mid$ $(i = 1, 2)$ are uniformly continuous on $[T_2, \infty)$. Using Barbalat's lemma (Gopalsamy 1992; Zhou et al. 2008):

$$\mid X_i(t) - \omega_i(t) \mid \to 0 \text{ as } t \to \infty, \ i = 1, 2$$

Hence, the theorem is proved. □

3.2.4 Prey–Predator System

In this section, we deal with the following prey–predator system, generated by the system (3.2.3) with (3.2.4):

$$\begin{aligned} \frac{dX_1(t)}{dt} &= X_1(t)[b_1(t) - b_{11}(t)X_1(t) - b_{12}(t)X_2(t)], \\ \frac{dX_2(t)}{dt} &= X_2(t)[-b_2(t) + b_{21}(t)X_1(t) - b_{22}(t)X_2(t)], \end{aligned} \tag{3.2.27}$$

where

$$b_1(t) = b_{10}(t) - r_1(t)\xi_0(t), \ b_2(t) = b_{20}(t) - r_2(t)\xi_0(t),$$

with initial conditions:

$$X_{10} = X_1(0) > 0, \ X_{20} = X_2(0) > 0. \tag{3.2.28}$$

The coefficients $b_i(t), b_{ij}(t)$ $(i, j = 1, 2)$ are assumed to be continuous and bounded on $[0, \infty)$ with $\min_{i,j=1,2}\{b_i^l, b_{ij}^l\} > 0$ and $\max_{i,j=1,2}\{b_i^u, b_{ij}^u\} < \infty$.
In this section, we analyse the permanence and global asymptotic stability of periodic solution of system (3.2.27) with (3.2.28).

Uniformly persistent

Theorem 3.9 *Let $X(t) = (X_1(t), X_2(t))^T$ be any solution of system (3.2.27) with (3.2.28). Let this system satisfy the following condition:*

$$b_{21}^u b_1^u > b_{11}^l b_2^l.$$

Then, $\exists\, T_2 > 0$ such that

$$X_i(t) \leq \Delta \ (i = 1, 2), \ \forall\, t \geq T_2, \text{ where } \Delta > \Delta^* = \max\left\{\frac{b_1^u}{b_{11}^l}, \frac{b_{21}^u b_1^u - b_{11}^l b_2^l}{b_{11}^l b_{22}^l}\right\}.$$

Proof Let $\Delta_1 > \frac{b_1^u}{b_{11}^l}$, we have, $\dot{X}_1(t) \leq X_1(t)[b_1(t) - b_{11}(t)X_1(t)] \leq X_1(t)[b_1^u - b_{11}^l X_1(t)]$.
Therefore, if $X_1(0) \leq \Delta_1$, then $X_1(t) \leq \Delta_1$, for all $t \geq 0$.

If $X_1(0) > \Delta_1$, and let $-\gamma_1 = \Delta_1(b_1^u - b_{11}^l\Delta_1)$, $\gamma_1 > 0$, then $\exists$ an $\epsilon_1 > 0$, s.t. if $t \in [0, \epsilon_1)$, $X_1(t) > \Delta_1$, and we have $\dot{X}_1(t) < -\gamma_1 < 0$.

Therefore, $\exists\ T_1 > 0$ s.t. $X_1(t) \leq \Delta_1,\ \forall\, t \geq T_1$.

Now, $\dot{X}_2(t) \leq X_2(t)[-b_2^l + b_{21}^u\Delta_1 - b_{22}^l X_2(t)],\ \forall\, t \geq T_1$. Here,

$$b_{21}^u\Delta_1 - b_2^l > b_{21}^u\frac{b_1^u}{b_{11}^l} - b_2^l = \frac{b_{21}^u b_1^u - b_{11}^l b_2^l}{b_{11}^l} > 0, \text{ by stated condition.}$$

Let $\Delta_2 > \dfrac{b_{21}^u\Delta_1 - b_2^l}{b_{22}^l}$, therefore, if $X_2(T_1) \leq \Delta_2$, then $X_2(t) \leq \Delta_2,\ \forall\, t \geq T_1$.

If $X_2(T_1) > \Delta_2$, then put $-\gamma_2 = \Delta_2[-b_2^l + b_{21}^u\Delta_1 - b_{22}^l\Delta_2]$, $\gamma_2 > 0$, then $\exists$ an $\epsilon_2 > 0$, s.t. if $t \in [T_1, T_1 + \epsilon_2)$, $X_2(t) > \Delta_2$, and we have $\dot{X}_2(t) < -\gamma_2 < 0$.

Therefore, $\exists\ T_2 > T_1 > 0$ s.t. $X_2(t) \leq \Delta_2,\ \forall\, t \geq T_2$.

Now, Δ_1 and Δ_2 can be chosen close enough to $\frac{b_1^u}{b_{11}^l}$ and $\frac{b_{21}^u\Delta_1 - b_2^l}{b_{22}^l}$, respectively. From the discussions, we conclude that $X_i(t) \leq \Delta\ (i = 1, 2),\ \forall\, t \geq T_2$, where

$$\Delta > \Delta^* = \max\left\{\frac{b_1^u}{b_{11}^l}, \frac{b_{21}^u b_1^u - b_{11}^l b_2^l}{b_{11}^l b_{22}^l}\right\}.$$

Hence, the theorem is proved. □

Theorem 3.10 *If the system (3.2.27) with (3.2.28) satisfies the following condition:*

$$b_{21}^u b_1^u > b_{11}^l b_2^l \textit{ and } \max\left\{\frac{b_1^u}{b_{11}^l}, \frac{b_{21}^u b_1^u - b_{11}^l b_2^l}{b_{11}^l b_{22}^l}\right\} < \min\left\{\frac{b_1^l}{b_{12}^u}, \frac{b_{21}^l b_1^l - b_{11}^u b_2^u}{b_{21}^l b_{12}^u}\right\},$$

then this system is uniformly persistent.

Proof Let $X(t) = (X_1(t), X_2(t))^{\mathrm{T}}$ be a solution of system (3.2.27) with (3.2.28).
Assume:

$$\max\left\{\frac{b_1^u}{b_{11}^l}, \frac{b_{21}^u b_1^u - b_{11}^l b_2^l}{b_{11}^l b_{22}^l}\right\} < \Delta < \min\left\{\frac{b_1^l}{b_{12}^u}, \frac{b_{21}^l b_1^l - b_{11}^u b_2^u}{b_{21}^l b_{12}^u}\right\}. \quad (3.2.29)$$

$\therefore \dot{X}_1(t) \geq X_1(t)[b_1^l - b_{11}^u X_1(t) - b_{12}^u \Delta]$, $\forall t \geq T_2$ (using (3.2.29) and Theorem (3.9)).
Now, by (3.2.29), $b_1^l - b_{12}^u \Delta > 0$. Let $0 < \zeta_1 < \frac{b_1^l - b_{12}^u \Delta}{b_{11}^u}$.
If $X_1(T_2) \geq \zeta_1$, then $X_1(t) \geq \zeta_1$, $\forall t \geq T_2$. If $X_1(T_2) < \zeta_1$, and let $\gamma_1 = X_1(T_2)(b_1^l - b_{11}^u \zeta_1 - b_{12}^u \Delta) > 0$, then $\exists$ an $\epsilon_1 > 0$, s.t.

$$X_1(t) < \zeta_1, \text{ and } \dot{X}_1(t) > \gamma_1 > 0, \ \forall t \in [T_2, T_2 + \epsilon_1).$$

Therefore, $\exists \, T_3 > T_2 > 0$, s.t. $X_1(t) \geq \zeta_1$, $\forall t \geq T_3$.
Also, $\dot{X}_2(t) \geq X_2(t)[-b_2^u + b_{21}^l X_1(t) - b_{22}^u X_2(t)] \geq X_2(t)[-b_2^u + b_{21}^l \zeta_1 - b_{22}^u X_2(t)]$, $\forall t \geq T_3$.
From (3.2.29), we have $\Delta < \frac{b_{21}^l b_1^l - b_{11}^u b_2^u}{b_{21}^l b_{12}^u} \Rightarrow \frac{b_2^u}{b_{21}^l} < \frac{b_1^l - b_{12}^u \Delta}{b_{11}^u}$. Choose ζ_1 in such a way that (without loss of generality),

$$\frac{b_2^u}{b_{21}^l} < \zeta_1 < \frac{b_1^l - b_{12}^u \Delta}{b_{11}^u} \Rightarrow b_{21}^l \zeta_1 - b_2^u > 0. \text{ Let } 0 < \zeta_2 < \frac{b_{21}^l \zeta_1 - b_2^u}{b_{22}^u}.$$

If $X_2(T_3) \geq \zeta_2$, then $X_2(t) \geq \zeta_2$, $\forall t \geq T_3$. If $X_2(T_3) < \zeta_2$, and let $\gamma_2 = X_2(T_3)(-b_2^u + b_{21}^l \zeta_1 - b_{22}^u \zeta_2) > 0$, then $\exists$ an $\epsilon_2 > 0$, s.t.

$$X_2(t) < \zeta_2, \text{ and } \dot{X}_2(t) > \gamma_2 > 0, \ \forall t \in [T_3, T_3 + \epsilon_2).$$

Therefore, $\exists \, T_4 > T_3 > T_2 > T_1 > 0$, s.t. $X_2(t) \geq \zeta_2$, $\forall t \geq T_4$.
From the discussions, we conclude that $\exists \, T_4 > T_3 > T_2 > T_1 > 0$ s.t. every solution of system (3.2.27) with (3.2.28) eventually enters and remains in the region $B = \{(X_1, X_2) \mid \zeta \leq X_i \leq \Delta, \ i = 1, 2\}$, $\forall t \geq T_4$, where $\zeta = \min\{\zeta_1, \zeta_2\}$.
Hence, the theorem is proved. □

Global asymptotic stability of periodic solution

Proceeding as in Theorem 3.8:

Theorem 3.11 *If the system (3.2.27) with (3.2.28) has a periodic solution, then this system is globally asymptotically stable provided for $i = 1, 2$:*

$$b_{21}^u b_1^u > b_{11}^l b_2^l, \ \max\left\{\frac{b_1^u}{b_{11}^l}, \frac{b_{21}^u b_1^u - b_{11}^l b_2^l}{b_{11}^l b_{22}^l}\right\} < \min\left\{\frac{b_1^l}{b_{12}^u}, \frac{b_{21}^l b_1^l - b_{11}^u b_2^u}{b_{21}^l b_{12}^u}\right\} \text{ and } b_{ii}^l > b_{3-i\,i}^u.$$

3.2.5 Cooperation System

In this section, we are going to study the following cooperation system generated by the system (3.2.3) with (3.2.4):

$$\frac{dX_1(t)}{dt} = X_1(t)[b_1(t) - b_{11}(t)X_1(t) + b_{12}(t)X_2(t)],$$
$$\frac{dX_2(t)}{dt} = X_2(t)[b_2(t) + b_{21}(t)X_1(t) - b_{22}(t)X_2(t)], \tag{3.2.30}$$

where

$$b_1(t) = b_{10}(t) - r_1(t)\xi_0(t),\ b_2(t) = b_{20}(t) - r_2(t)\xi_0(t),$$

with initial conditions:

$$X_{10} = X_1(0) > 0,\ X_{20} = X_2(0) > 0. \tag{3.2.31}$$

The coefficients $b_i(t), b_{ij}(t)$ $(i, j = 1, 2)$ are assumed to be continuous and bounded on $[0, \infty)$ with $\min_{i,j=1,2}\{b_i^l, b_{ij}^l\} > 0$ and $\max_{i,j=1,2}\{b_i^u, b_{ij}^u\} < \infty$.
In this section, we analyse the permanence and global asymptotic stability of periodic solution of system (3.2.30) with (3.2.31).

Uniformly persistent

Theorem 3.12 *Let $X(t) = (X_1(t), X_2(t))^T$ be any solution of system (3.2.30) with (3.2.31). Suppose this system satisfies the following conditions:*

$$b_{ii}^l > b_{i\,3-i}^u\ (i = 1, 2),\ \textit{then it is uniformly persistent.}$$

Proof Let us define $\Omega(t) = \max\{X_1(t), X_2(t)\}$.
Calculating the upper-right derivative of $\Omega(t)$ along the solution of system (3.2.30) with (3.2.31), we have the following:

(1) If $X_1(t) \geq X_2(t)$, then
$$D^+\Omega(t) = \dot{X}_1(t) \leq X_1(t)[b_1^u - b_{11}^l X_1(t) + b_{12}^u X_2(t)]$$
$$\leq X_1(t)[b_1^u - (b_{11}^l - b_{12}^u)X_1(t)]. \tag{3.2.32}$$

(2) If $X_1(t) \leq X_2(t)$, then
$$D^+\Omega(t) = \dot{X}_2(t) \leq X_2(t)[b_2^u + b_{21}^u X_1(t) - b_{22}^l X_2(t)]$$
$$\leq X_2(t)[b_2^u - (b_{22}^l - b_{21}^u)X_2(t)]. \tag{3.2.33}$$

Let $\Delta > \Delta^* = \max_{i=1,2}\left\{\frac{b_i^u}{b_{ii}^l - b_{i\,3-i}^u}\right\} > 0$, since by stated conditions $b_{ii}^l - b_{i\,3-i}^u > 0$, $(i = 1, 2)$. From (3.2.32) and (3.2.33), we can get the following:

(i) If $\max\{X_1(0), X_2(0)\} \leq \Delta$, then $\max\{X_1(t), X_2(t)\} \leq \Delta,\ \forall\, t \geq 0$.

(ii) If $\max\{X_1(0), X_2(0)\} > \Delta$, and let $-\gamma_1 = \max_{i=1,2}[\Delta\{b_i^u - (b_{ii}^l - b_{i\,3-i}^u)\Delta\}]$, $\gamma_1 > 0$, we have the following three possibilities:

$$\text{(a) } \Omega(0) = X_1(0) > \Delta, \ (X_1(0) > X_2(0));$$

$$\text{(b) } \Omega(0) = X_2(0) > \Delta, \ (X_1(0) < X_2(0));$$

$$\text{(c) } \Omega(0) = X_1(0) = X_2(0) > \Delta.$$

If (a) holds, then $\exists$ an $\epsilon > 0$, s.t. if $t \in [0, \epsilon)$, $\Omega(t) = X_1(t) > \Delta$, and we have $D^+\Omega(t) = \dot{X}_1(t) < -\gamma_1 < 0$.
If (b) holds, then $\exists$ an $\epsilon > 0$, s.t. if $t \in [0, \epsilon)$, $\Omega(t) = X_2(t) > \Delta$, and we have $D^+\Omega(t) = \dot{X}_2(t) < -\gamma_1 < 0$.
If (c) holds, then $\exists$ an $\epsilon > 0$, s.t. if $t \in [0, \epsilon)$, $\Omega(t) = X_i(t) > \Delta$, $(i = 1, 2)$. Similar to (a) and (b), we have $D^+\Omega(t) = \dot{X}_i(t) < -\gamma_1 < 0$ $(i = 1, 2)$.
From investigating (a), (b) and (c), we conclude that if $\Omega(0) > \Delta$, then $\exists\, T_1 > 0$ s.t.

$$\Omega(t) = \max\{X_1(t), X_2(t)\} \le \Delta, \forall\, t \ge T_1$$

Let $X(t) = (X_1(t), X_2(t))^{\mathrm{T}}$ be a solution of (3.2.30) and (3.2.31). Therefore,

$$\dot{X}_1(t) \ge X_1(t)[b_1^l - b_{11}^u X_1(t) - b_{12}^u \Delta], \ \forall\, t \ge T_1,$$

$$\dot{X}_2(t) \ge X_2(t)[b_2^l - b_{22}^u X_2(t) - b_{21}^u \Delta], \ \forall\, t \ge T_1.$$

From the stated conditions, we have $b_i^l - b_{i\,3-i}^u \Delta > 0, \ i = 1, 2$.

Define $\Omega_1(t) = \min\{X_1(t), X_2(t)\}$. Then, calculating the lower-right derivative of $\Omega_1(t)$ along the solution of system (3.2.30) with (3.2.31), we get:

$$\text{(3) If } X_1(t) \le X_2(t), \text{ then } D_+\Omega_1(t) = \dot{X}_1(t) \ge X_1(t)[b_1^l - b_{11}^u X_1(t)], \ \forall\, t \ge 0. \tag{3.2.34}$$

$$\text{(4) If } X_1(t) \ge X_2(t), \text{ then } D_+\Omega_1(t) = \dot{X}_2(t) \ge X_2(t)[b_2^l - b_{22}^u X_2(t)], \ \forall\, t \ge 0. \tag{3.2.35}$$

Let $\zeta^* = \min\left\{\frac{b_1^l}{b_{11}^u}, \frac{b_2^l}{b_{22}^u}\right\}$, we choose ζ as $0 < \zeta < \zeta^*$. From (3.2.34) and (3.2.35), we conclude the following:

(i) If $\Omega_1(0) = \min\{X_1(0), X_2(0)\} \ge \zeta$, then $\min\{X_1(t), X_2(t)\} \ge \zeta, \ \forall\, t \ge 0$.

(ii) If $\Omega_1(0) = \min\{X_1(0), X_2(0)\} < \zeta$, and let $\gamma_2 = \min_{i=1,2}\{X_i(0)(b_i^l - b_{ii}^u \zeta)\}$, we consider the following three cases:

$$\text{(d) } \Omega_1(0) = X_1(0) < \zeta, \ (X_1(0) < X_2(0));$$

$$\text{(e) } \Omega_1(0) = X_2(0) < \zeta, \ (X_2(0) < X_1(0));$$

$$\text{(f) } \Omega_1(0) = X_1(0) = X_2(0) < \zeta.$$

If (d) holds, then $\exists$ an $\epsilon > 0$, s.t. if $t \in [0, \epsilon)$, $\Omega_1(t) = X_1(t) < \zeta$, and we have $D_+\Omega_1(t) = \dot{X}_1(t) > \gamma_2 > 0$.
If (e) holds, then $\exists$ an $\epsilon > 0$, s.t. if $t \in [0, \epsilon)$, $\Omega_1(t) = X_2(t) < \zeta$, and we have $D_+\Omega_1(t) = \dot{X}_2(t) > \gamma_2 > 0$.
If (f) holds, then $\exists$ an $\epsilon > 0$, s.t. if $t \in [0, \epsilon)$, $\Omega_1(t) = X_i(t) < \zeta$, $(i = 1, 2)$. Similar to (d) and (e), we have $D_+\Omega_1(t) = \dot{X}_i(t) > \gamma_2 > 0$ $(i = 1, 2)$.
From investigating (d), (e) and (f), we conclude that if $\Omega_1(0) = \min\{X_1(0), X_2(0)\} < \zeta$, then $\exists\, T_2 > 0$ s.t.

$$\Omega_1(t) = \min\{X_1(t), X_2(t)\} \geq \zeta, \forall\, t \geq T_2.$$

From the discussions, we conclude that $\exists\, T = \max(T_1, T_2) > 0$ s.t. every solution of system (3.2.30) with (3.2.31) eventually enters and remains in the region $B = \{(X_1, X_2) \mid \zeta \leq X_i \leq \Delta,\ i = 1, 2\},\ \forall\, t \geq T$. Hence, the theorem is proved. □

Global asymptotic stability of periodic solution

Proceeding as in Theorem 3.8:

Theorem 3.13 *If the system (3.2.30) with (3.2.31) has a periodic solution, then this system is globally asymptotically stable provided*

$$b^l_{ii} > \max\{b^u_{i\,3-i}, b^u_{3-i\,i}\} \ (i = 1, 2).$$

3.2.6 Discussion

In this article, we have studied the effect of toxicants on a two-species nonautonomous Lotka–Volterra model. We have focused the global asymptotic behaviour of this model by developing bounded functions $H_i > 0$ $(i = 1, 2)$, satisfying equations which are suitable perturbations of original stated equations. The survival threshold of each population for competition, predation and cooperation systems, respectively, has been obtained. By constructing a suitable Lyapunov function, we have obtained sufficient conditions to confirm that if each of competition, predation and cooperation systems, respectively, admits a positive periodic solution, then it is globally asymptotically stable. Our findings divulge the fact that lower value of the dose response parameter of prey species to the organismal toxicant concentration at time t leads to make the prey–predator system uniformly persistent. But the toxicants have no effect on uniformly persistent for cooperation system.

3.3 Analysis of a Nonautonomous Delayed Prey–Predator System with a Stage Structure for the Predator in a Polluted Environment

In this natural world, there are so many species whose individual members have a life history that takes them through two stages: immature and mature. Moreover, it is observed that there is a huge difference between the mature and immature species in many aspects. As for example, for less concentration of C_2H_4 in an environment, the immature *cymbidium* is so susceptible that it starts to wither, whereas the mature *cymbidium* usually has not been affected (Manning and Feder 1980; Xiao and Chen 2001). Therefore, it is very important to take into account the effect of stage structure on population growth in a polluted environment. In this article, we have considered pollution in a nonautonomous prey–predator system together with diffusional migration among the immature predator population between protective and nonprotective patches where there is a constant delay among the matured predator due to the gestation of the predator subject to the assumption that mature adult predator can only contribute to the reproduction of the predator biomass and that the change rate of predator depends on the number of the prey and of the predator presented at some previous time. The scale or degree of the protective zone among the immature predator population can be controlled through diffusive coefficients $\delta_i(t),\ (i = 1, 2)$. In this study, we have incorporated stage structure for the predator and assumed that the mature species is free from the effects toxicants (seems to be reasonable). The main purpose of the present study is to analyse the asymptotic behaviour of the toxicant population model taking into account the stage structure, diffusional migration and time delay so as to derive some conditions under which the population is uniformly persistent (means the survival of all species which exist initially). In addition, by constructing an appropriate function (called Lyapunov function), we have got sufficient conditions to confirm that if the system admits a positive periodic solution, then it is globally asymptotically stable.

3.3.1 The Basic Mathematical Model

This study is based on the hypothesis of a total spatial homogeneous environment with stage structure for the predator into a nonautonomous Lotka–Volterra-type prey–predator model. The following delayed model is considered to describe the dynamics of population of two species in a polluted environment where the state variables $X(t)$, $Y_1(t)$, $Y_2(t)$ and $Y_3(t)$ represent the densities of the prey, immature predator in nonprotective patch (with toxicants), immature predator in protective patch (without toxicants) and mature predator species at time t, respectively, $\xi_0(t)$ and $\xi_e(t)$ stand for the concentrations of toxicant in the organism and in the environment at time t, respectively. It is further assumed that the toxicants have no effect on the mature species. The following model is considered with diffusional migration among the immature predator population, constant delay among the matured predator and toxicant effect on immature predator in nonprotective patch:

$$
\begin{aligned}
\frac{dX(t)}{dt} &= X(t)[b_1(t) - b_{11}(t)X(t) - b_{12}(t)Y_3(t)],\\
\frac{dY_1(t)}{dt} &= r_1(t)Y_3(t) - p_1(t)Y_1(t) - \lambda_1(t)Y_1(t) - \mathrm{d}(t)\xi_0(t)Y_1(t)\\
&\quad -\beta_1(t)Y_1^2(t) + \delta_1(t)(Y_2(t) - Y_1(t)),\\
\frac{dY_2(t)}{dt} &= r_2(t)Y_3(t) - p_2(t)Y_2(t) - \lambda_2(t)Y_2(t) - \beta_2(t)Y_2^2(t)\\
&\quad +\delta_2(t)(Y_1(t) - Y_2(t)),\\
\frac{dY_3(t)}{dt} &= Y_3(t)[-b_2(t) - b_{22}(t)Y_3(t)] + b_{21}(t)Y_3(t-\tau)X(t-\tau)\\
&\quad +p_1(t)Y_1(t) + p_2(t)Y_2(t),\\
\frac{d\xi_0(t)}{dt} &= k(t)\xi_e(t) + \rho(t) - g(t)\xi_0(t) - \zeta(t)\xi_0(t),\\
\frac{d\xi_e(t)}{dt} &= -\psi(t)\xi_e(t) + \eta(t)
\end{aligned}
\tag{3.3.1}
$$

Additional assumptions:

1. *Prey population*: (i) The growth of the species is of Lotka–Volterra type with birth rate at any time $t > 0$ is proportional to the existing prey population having proportional function $b_1(t)$.
   (ii) $b_{11}(t)$ stands for the intraspecific competition rate function of the prey.
   (iii) The functions $b_1(t)$ and $b_{11}(t)$ are continuous, bounded and strictly positive on $[0, \infty)$.
2. *Predator population*: (i) Prey cannot be killed by infant (immature) predator. The mature predator feed only on prey: $b_{12}(t)$ is the capturing rate function of the predator, the conversion rate function of nutrients into the reproduction of the mature predator is $\frac{b_{21}(t)}{b_{12}(t)}$.
   (ii) $\tau \geq 0$ is a constant delay because of the gestation of the mature predator based on the assumption that mature adult predator can only contribute to the reproduction of the predator biomass and that the change rate of predator depends on the densities of the prey and of the predator presented at some previous time.
   (iii) To secure the immature predator population, the region (Υ) of immature predator population is divided into two patches: Υ_1 and Υ_2 where pollution is allowed in Υ_1 (nonprotective) and is inhibited in Υ_2 (protective).
   (iv) Because of the difference of densities in Υ_1 and Υ_2, the diffusive migration can occur between Υ_1 and Υ_2 with respective diffusion coefficient functions $\delta_1(t)$ and $\delta_2(t)$. The rates of transitions from individuals in Υ_1 and Υ_2 to mature

individuals are proportional to the existing populations in Υ_1 and Υ_2, respectively, with respective proportional functions $p_1(t)$ and $p_2(t)$ at any time $t > 0$.
(v) The death rates of mature population, nonprotective immature population (in Υ_1) and protective immature population (in Υ_2) are proportional to the existing mature, nonprotective immature and protective immature population, respectively, with proportional functions $b_2(t)$, $\lambda_1(t)$ and $\lambda_2(t)$, respectively. The birth rates of nonprotective immature population (in Υ_1) and protective immature population (in Υ_2) are proportional to the existing mature population with proportional functions $r_1(t)$ and $r_2(t)$, respectively. $b_{22}(t)$, $\beta_1(t)$ and $\beta_2(t)$ are the intraspecific competition rate functions of the mature population, nonprotective immature population and protective immature population, respectively.
(vi) $b_{12}(t)$, $b_{21}(t)$, $b_{22}(t)$, $\delta_i(t)$, $p_i(t)$, $b_2(t)$, $\lambda_i(t)$, $r_i(t)$ and $\beta_i(t)$, $(i = 1, 2)$ are continuous, bounded and strictly positive functions on $[0, \infty)$.

3. *Toxicants*: $d(t)$ is the response parameter of nonprotective immature predator species $Y_1(t)$ (living in Υ_1) to the organismal toxicant concentration. The organismal net uptake of toxicant from the environment $(k(t)\xi_e(t))$ and the food chain $(\rho(t))$ are denoted in the first two terms on the right of the third equation in system (3.3.1), respectively, the organismal net loss of toxicant due to metabolic processing and other causes are represented by the third and fourth terms. Environmental toxicant uptake rate per unit mass organism is denoted by $k(t)$, uptake rate of toxicant in food per unit mass organism is denoted by $\rho(t)$, and the egestion and depuration rates of the toxicant in the organism are represented by $-g(t)\xi_0(t)$ and $-\zeta(t)\xi_0(t)$, respectively. The first term $(-\psi(t)\xi_e(t))$ on the right of the fourth equation in system (3.3.1) represents the toxicant loss from the environment itself by biological transformation, chemical hydrolysis, volatilization, microbial degradation, photosynthetic degradation and so on and $\eta(t)$ represents the exogenous rate of input of toxicant into the environment. The change of toxicants in the environment that comes from uptake and egestion by the organisms can be neglected due to the huge capacity of the environment. The coefficients $d(t), k(t), \rho(t), g(t), \zeta(t), \psi(t)$ and $\eta(t)$ are considered to be non-negative, continuous and bounded functions on $[0, \infty)$.

The initial conditions are as follows:

$$X(\varsigma) = \phi_1(\varsigma),\ Y_i(\varsigma) = \theta_i(\varsigma),\ \xi_0(\varsigma) = \phi_2(\varsigma),\ \xi_e(\varsigma) = \phi_3(\varsigma),\ (i = 1, 2, 3)$$

$$\phi_i(\varsigma) \geq 0,\ \psi_i(\varsigma) \geq 0,\ (i = 1, 2, 3),\ \varsigma \in [-\tau, 0]$$

$$X(0) > 0,\ Y_i(0) > 0,\ 0 \leq \xi_0(0) \leq 1,\ 0 \leq \xi_e(0) \leq 1,\ (i = 1, 2, 3), \tag{3.3.2}$$

where $\Phi = (\phi_1(\varsigma), \phi_2(\varsigma), \phi_3(\varsigma), \theta_1(\varsigma), \theta_2(\varsigma), \theta_3(\varsigma)) \in \mathbb{C}([-\tau, 0], R^6_+)$, the Banach space of continuous functions mapping the interval $[-\tau, 0]$ into R^6_+, where $R^6_+ = \{(X_1, X_2, X_3, X_4, X_5, X_6) : X_i \geq 0,\ i = 1, 2, 3, 4, 5, 6\}$.

Applying the fundamental theory of functional differential equations (Hale 1977), system (3.3.1) has a unique solution $(X(t), Y_1(t), Y_2(t), Y_3(t), \xi_0(t), \xi_e(t))$ satisfying initial conditions (3.3.2).
Let $\phi^l = \inf_{t\geq 0}\phi(t),\ \phi^u = \sup_{t\geq 0}\phi(t)$, for a continuous and bounded function $\phi(t)$ defined in the interval $[0, +\infty)$.
Assumptions:

$$\min\{d^l, k^l, \rho^l, g^l, \zeta^l, \psi^l, \eta^l\} \geq 0, \max\{d^u, k^u, \rho^u, g^u, \zeta^u, \psi^u, \eta^u\} < \infty,$$

$$\min_{i,j=1,2}\{b_i^l, b_{ij}^l, r_i^l, p_i^l, \lambda_i^l, \beta_i^l, \delta_i^l\} > 0,\ \max_{i,j=1,2}\{b_i^u, b_{ij}^u, r_i^u, p_i^u, \lambda_i^u, \beta_i^u, \delta_i^u\} < \infty. \tag{3.3.3}$$

Each of $\xi_0(t)$ and $\xi_e(t)$ is a concentration, and so these variables cannot be greater than 1; that is, some conditions should be needed such that $0 \leq \xi_0(t) \leq 1,\ 0 \leq \xi_e(t) \leq 1$, for all $t \geq 0$.

Theorem 3.14 *Every solution of system (3.3.1) with initial conditions (3.3.2) uniquely exists and* $X(t) > 0,\ Y_1(t) > 0,\ Y_2(t) > 0,\ Y_3(t) > 0,\ \xi_0(t) \geq 0,\ \xi_e(t) \geq 0$, *for all* $t \in [0, \infty)$.

Proof As the right-hand side of system (3.3.1) is completely continuous and locally Lipschitzian on $\mathbb{C}$, the solution $(X(t), Y_1(t), Y_2(t), Y_3(t), \xi_0(t), \xi_e(t))$ of (3.3.1) with initial conditions (3.3.2) exists and is unique on $[0, \beta)$, where $0 < \beta \leq +\infty$ (Hale 1977; Chap. 2). From system (3.3.1) with (3.3.2):

$$X(t) = X(0)\exp\int_0^t [b_1(s) - b_{11}(s)X(s) - b_{12}(s)Y_3(s)]\mathrm{d}s > 0,\ \forall\, t \geq 0.$$

Next, let us prove that $Y_1(t) > 0$ on $t \in [0, \infty)$. Otherwise, $\exists\ t_1 \in (0, \infty)$ such that $Y_1(t_1) = 0,\ \dot{Y}_1(t_1) \leq 0$ and $Y_1(t) > 0,\ \forall\, t \in [0, t_1)$. Hence, it must have $Y_2(t) > 0,\ \forall\, t \in [0, t_1)$. If it is not true, then $\exists\ t_2 \in (0, t_1)$ such that $Y_2(t_2) = 0$ and $Y_2(t) > 0$ on $[0, t_2)$. We claim that $Y_3(t) > 0\ \forall\ t \in [0, t_2)$. If this statement is false, then $\exists\ t_3 \in (0, t_2)$ such that $Y_3(t_3) = 0$ and $Y_3(t) \geq 0,\ \forall\, t \in [-\tau, t_3]$. Furthermore:

$$\frac{\mathrm{d}Y_3(t)}{\mathrm{d}t} \geq -b_2(t)Y_3(t) - b_{22}(t)Y_3^2(t),\ \text{ for all } t \in [0, t_3].$$

$$\text{Then, }\ Y_3(t) \geq Y_3(0)\exp\int_0^t [-b_2(s) - b_{22}(s)Y_3(s)]\mathrm{d}s > 0,\ \forall\, t \in [0, t_3].$$

$$\Rightarrow Y_3(t_3) \geq Y_3(0)\exp\int_0^{t_3} [-b_2(s) - b_{22}(s)Y_3(s)]\mathrm{d}s > 0,$$

which is a contradiction with $Y_3(t_3) = 0$. Hence, $Y_3(t) > 0,\ \forall\, t \in [0, t_2)$. Integrating the third equation of system (3.3.1) from 0 to t_2, we have:

$$Y_2(t_2) = Y_2(0)\exp\left\{-\int_0^{t_2}(\delta_2(s) + p_2(s) + \lambda_2(s) + \beta_2(s)Y_2(s))\mathrm{d}s\right\} + \int_0^{t_2}(r_2(u)Y_3(u) + \delta_2(u)Y_1(u)) \exp\left\{\int_{t_2}^{u}(\delta_2(s) + p_2(s) + \lambda_2(s) + \beta_2(s)Y_2(s))\mathrm{d}s\right\}\mathrm{d}u > 0,$$

which is a contradiction. So $Y_2(t) > 0,\ \forall\, t \in [0, t_1)$ and hence $Y_3(t) > 0,\ \forall\, t \in [0, t_1)$. Integrating the second equation of system (3.3.1) from 0 to t_1:

$$Y_1(t_1) = Y_1(0)\exp\left\{-\int_0^{t_1}(\delta_1(s) + p_1(s) + \lambda_1(s) + \mathrm{d}(s)\xi_0(s) + \beta_1(s)Y_1(s))\mathrm{d}s\right\} + \int_0^{t_1}(r_1(u)Y_3(u) + \delta_1(u)Y_2(u))\exp\{\int_{t_1}^{u}(\delta_1(s) + p_1(s) + \lambda_1(s) + d(s)\xi_0(s) + \beta_1(s)Y_1(s))\mathrm{d}s\}\mathrm{d}u > 0,$$

a contradiction with $Y_1(t_1) = 0$. So $Y_1(t) > 0,\ \forall t \geq 0$ and hence $Y_2(t) > 0,\ Y_3(t) > 0,\ \forall\, t \geq 0$.

$$\text{Also, } \xi_0(t) = \mathrm{e}^{-\int_0^t\{\zeta(s)+g(s)\}\mathrm{d}s}[\int_0^t \{k(s)\xi_\mathrm{e}(s) + \rho(s)\}\mathrm{e}^{\int_0^s\{\zeta(s)+g(s)\}\mathrm{d}s}\mathrm{d}s + \xi_0(0)] \geq 0,$$

$$\text{and, } \xi_\mathrm{e}(t) = \mathrm{e}^{-\int_0^t\psi(s)\mathrm{d}s}[\int_0^t \eta(s)\mathrm{e}^{\int_0^s\psi(s)\mathrm{d}s}\mathrm{d}s + \xi_\mathrm{e}(0)] \geq 0, \text{ for all } t \geq 0.$$

Hence, the theorem is proved. □

Theorem 3.15 *From (3.3.1) with (3.3.2): if* $k^u + \rho^u \leq g^l + \zeta^l,\ \eta^u \leq \psi^l$, *then* $0 \leq \xi_0(t) \leq 1,\ 0 \leq \xi_e(t) \leq 1$, *for all* $t \geq 0$.

Proof Proceed as in the proof of Theorem 3.2 of Sect. 3.2.1. □

Substituting the expression of $\xi_e(t)$ in $\xi_0(t)$, we can express $\xi_0(t)$ in term of some bounded continuous functions. Therefore, the system (3.3.1) with (3.3.2) may be simplified as follows:

$$\begin{aligned}
\frac{dX(t)}{dt} &= X(t)[b_1(t) - b_{11}(t)X(t) - b_{12}(t)Y_3(t)],\\
\frac{dY_1(t)}{dt} &= r_1(t)Y_3(t) - p_1(t)Y_1(t) - \lambda_1(t)Y_1(t) - d(t)\xi_0(t)Y_1(t)\\
&\quad -\beta_1(t)Y_1^2(t) + \delta_1(t)(Y_2(t) - Y_1(t)),\\
\frac{dY_2(t)}{dt} &= r_2(t)Y_3(t) - p_2(t)Y_2(t) - \lambda_2(t)Y_2(t) - \beta_2(t)Y_2^2(t)\\
&\quad +\delta_2(t)(Y_1(t) - Y_2(t)),\\
\frac{dY_3(t)}{dt} &= Y_3(t)[-b_2(t) - b_{22}(t)Y_3(t)] + b_{21}(t)Y_3(t-\tau)X(t-\tau)\\
&\quad +p_1(t)Y_1(t) + p_2(t)Y_2(t),
\end{aligned} \tag{3.3.4}$$

with initial conditions are:

$$\begin{aligned}
&X(\varsigma) = \phi_1(\varsigma),\ Y_i(\varsigma) = \theta_i(\varsigma),\ (i = 1, 2, 3)\\
&\phi_1(\varsigma) \geq 0,\ \theta_i(\varsigma) \geq 0,\ (i = 1, 2, 3),\ \theta \in [-\tau, 0]\\
&X(0) > 0,\ Y_i(0) > 0,\ (i = 1, 2, 3),
\end{aligned} \tag{3.3.5}$$

where $\Phi = (\phi_1(\varsigma), \theta_1(\varsigma), \theta_2(\varsigma), \theta_3(\varsigma)) \in \mathbb{C}([-\tau, 0], R_+^4)$, the Banach space of continuous functions maps the interval $[-\tau, 0]$ into R_+^4, where $R_+^4 = \{(X_1, X_2, X_3, X_4) : X_i \geq 0,\ i = 1, 2, 3, 4\}$.

3.3.2 *Uniformly Persistent of System (3.3.4)*

Here, we would like to discuss the uniformly persistent of the system (3.3.4) with (3.3.5), which demonstrates how this system will survive forever (i.e. will not be going to extinct in time), under some conditions.

Definition The system (3.3.4) is said to be uniformly persistent, in the ecological context, it implies the survival of all species which exist initially. That is, the long-term survival (will not vanish in time) of all components of the system (3.3.4), if there are positive constants ν_i and σ_i $(i = 1, 2, 3, 4)$ such that:

$$\nu_1 \leq \liminf_{t\to\infty} X(t) \leq \limsup_{t\to\infty} X(t) \leq \sigma_1,$$

$$\nu_2 \leq \liminf_{t\to\infty} Y_1(t) \leq \limsup_{t\to\infty} Y_1(t) \leq \sigma_2,$$

$$\nu_3 \leq \liminf_{t\to\infty} Y_2(t) \leq \limsup_{t\to\infty} Y_2(t) \leq \sigma_3,$$

$$\nu_4 \leq \liminf_{t\to\infty} Y_3(t) \leq \limsup_{t\to\infty} Y_3(t) \leq \sigma_4,$$

hold for any solution $(X(t), Y_1(t), Y_2(t), Y_3(t))$ of (3.3.4) with (3.3.5). Here, ν_i and σ_i $(i = 1, 2, 3, 4)$ are independent of the initial conditions (3.3.5).

Theorem 3.16 *(Song and Chen 2001). Consider the following ordinary delay differential equation:*

$$\dot{X}(t) = a_1 X(t-\tau) - a_2 X(t) - a_3 X^2(t),$$

where $a_1, a_2, a_3, \tau > 0$ with $X(t) > 0$, for $-\tau \leq t \leq 0$. We have

(I) if $a_1 > a_2$, then $\lim_{t\to\infty} X(t) = \dfrac{a_1 - a_2}{a_3}$; and (II) if $a_1 < a_2$, then $\lim_{t\to\infty} X(t) = 0$.

Theorem 3.17 *Let $(X(t), Y_1(t), Y_2(t), Y_3(t))^T$ be any solution of the system (3.3.4) with (3.3.5) and assume the following: $r_1^u > p_1^l + \lambda_1^l + d^l \xi_0^l$, $r_2^u > p_2^l + \lambda_2^l$ and $b_{21}^u b_1^u > b_{11}^l (b_2^l - p_1^u - p_2^u) > 0$. Then, $\exists\, T_3 > 0$ such that*

$$X(t) \leq \Delta_1 \text{ and } Y_i(t) \leq \Delta_4 \ (i = 1, 2, 3),\ \forall\, t \geq T_3, \text{ where } \Delta_1 > \frac{b_1^u}{b_{11}^l} \text{ and}$$

$$\Delta_4 > \Delta^* = \max\left\{\frac{r_1^u - p_1^l - \lambda_1^l - d^l \xi_0^l}{\beta_1^l}, \frac{r_2^u - p_2^l - \lambda_2^l}{\beta_2^l}, \frac{b_{21}^u b_1^u - b_{11}^l (b_2^l - p_1^u - p_2^u)}{b_{11}^l b_{22}^l}\right\}.$$

Proof Let $\Delta_1 > \frac{b_1^u}{b_{11}^l}$. We have, $\dot{X}(t) \leq X(t)[b_1(t) - b_{11}(t)X(t)] \leq X(t)[b_1^u - b_{11}^l X(t)]$.
Therefore, if $X(0) \leq \Delta_1$, then $X(t) \leq \Delta_1$, $\forall\, t \geq 0$.

If $X(0) > \Delta_1$, and let $-\gamma_1 = \Delta_1(b_1^u - b_{11}^l \Delta_1)$, $\gamma_1 > 0$, then $\exists$ an $\epsilon_1 > 0$,

s.t. if $t \in [0, \epsilon_1)$, $X(t) > \Delta_1$, and we have $\dot{X}(t) < -\gamma_1 < 0$.

Therefore, $\exists\, T_1 > 0$ s.t. $X(t) \leq \Delta_1$, $\forall\, t \geq T_1$, where $\Delta_1 > \dfrac{b_1^u}{b_{11}^l}$.

We define $\Omega(t) = \max\{Y_1(t), Y_2(t), Y_3(t)\}$.

Calculating the upper-right derivative of $\Omega(t)$ along the solution of system (3.3.4) with (3.3.5), we have the following:

$$(1)\text{ If } Y_1(t) \geq Y_2(t) \geq Y_3(t), \text{ or } Y_1(t) \geq Y_3(t) \geq Y_2(t), \text{ then } D^+\Omega(t) = \dot{Y}_1(t)$$

$$\leq r_1^u Y_1(t) - p_1^l Y_1(t) - \lambda_1^l Y_1(t) - \mathrm{d}^l \xi_0^l Y_1(t) - \beta_1^l Y_1^2(t)$$

$$= (r_1^u - p_1^l - \lambda_1^l - \mathrm{d}^l \xi_0^l) Y_1(t) - \beta_1^l Y_1^2(t), \ \forall t \geq 0. \tag{3.3.6}$$

$$(2)\text{ If } Y_2(t) \geq Y_1(t) \geq Y_3(t), \text{ or } Y_2(t) \geq Y_3(t) \geq Y_1(t), \text{ then } D^+\Omega(t) = \dot{Y}_2(t)$$

$$\leq r_2^u Y_2(t) - p_2^l Y_2(t) - \lambda_2^l Y_2(t) - \beta_2^l Y_2^2(t)$$

$$= (r_2^u - p_2^l - \lambda_2^l) Y_2(t) - \beta_2^l Y_2^2(t), \ \forall t \geq 0. \tag{3.3.7}$$

$$(3)\text{ If } Y_3(t) \geq Y_1(t) \geq Y_2(t), \text{ or } Y_3(t) \geq Y_2(t) \geq Y_1(t), \text{ then } D^+\Omega(t) = \dot{Y}_3(t)$$

$$\leq -b_2^l Y_3(t) - b_{22}^l Y_3^2(t) + b_{21}^u \Delta_1 Y_3(t-\tau) + (p_1^u + p_2^u) Y_3(t)$$

$$= -(b_2^l - p_1^u - p_2^u) Y_3(t) - b_{22}^l Y_3^2(t) + b_{21}^u \Delta_1 Y_3(t-\tau), \ \forall t \geq T_1 + \tau. \tag{3.3.8}$$

$$\text{Let, } \Delta_2 > \Delta_1^* = \max\left\{ \frac{r_1^u - p_1^l - \lambda_1^l - d^l \xi_0^l}{\beta_1^l}, \frac{r_2^u - p_2^l - \lambda_2^l}{\beta_2^l} \right\}.$$

From (3.3.6) and (3.3.7), we can get the following:

(i) If $\max\{Y_1(0), Y_2(0), Y_3(0)\} \leq \Delta_2$, then $\max\{Y_1(t), Y_2(t), Y_t(t)\} \leq \Delta_2, \ \forall t \geq 0$.
(ii) If $\max\{Y_1(0), Y_2(0), Y_3(0)\} > \Delta_2$, and let $-\gamma_2 = \max[\Delta_2\{(r_1^u - p_1^l - \lambda_1^l - d^l \xi_0^l) - \beta_1^l \Delta_2\}$,

$\Delta_2\{(r_2^u - p_2^l - \lambda_2^l) - \beta_2^l \Delta_2\}], \gamma_2 > 0$, we consider the following two cases:

$$\text{(a) } \Omega(0) = Y_1(0) > \Delta_2,$$

$$\text{(b) } \Omega(0) = Y_2(0) > \Delta_2.$$

If (a) holds, then $\exists\, \epsilon_2 > 0$, s.t. if $t \in [0, \epsilon_2)$, $\Omega(t) = Y_1(t) > \Delta_2$, and we have $D^+\Omega(t) = \dot{Y}_1(t) < -\gamma_2 < 0$.
If (b) holds, then $\exists\, \epsilon_3 > 0$, s.t. if $t \in [0, \epsilon_3)$, $\Omega(t) = Y_2(t) > \Delta_2$, and we have $D^+\Omega(t) = \dot{Y}_2(t) < -\gamma_2 < 0$.

If (3.3.8) holds, then by Theorem 3.16, $\exists\, T_2 \geq T_1 + \tau > 0$, s.t.

$$Y_3(t) < \Delta_3,\ \forall\, t \geq T_2 \geq T_1 + \tau > 0,\ \text{where } \Delta_3 > \frac{b_{21}^u b_1^u - b_{11}^l (b_2^l - p_1^u - p_2^u)}{b_{11}^l b_{22}^l}$$

$$\left(\text{since } \Delta_1 \text{ can be chosen close enough to } \frac{b_1^u}{b_{11}^l}\right).$$

From the discussions, we can conclude that $\exists\, T_3 \geq T_2 \geq T_1 + \tau > 0$, s.t.

$$X(t) \leq \Delta_1 \text{ and } Y_i(t) \leq \Delta_4\ (i = 1, 2, 3),\ \forall\, t \geq T_3,\ \text{where } \Delta_1 > \frac{b_1^u}{b_{11}^l} \text{ and}$$

$$\Delta_4 > \Delta^* = \max\left\{\frac{r_1^u - p_1^l - \lambda_1^l - d^l \xi_0^l}{\beta_1^l}, \frac{r_2^u - p_2^l - \lambda_2^l}{\beta_2^l}, \frac{b_{21}^u b_1^u - b_{11}^l (b_2^l - p_1^u - p_2^u)}{b_{11}^l b_{22}^l}\right\}.$$

Hence, the theorem is proved. □

Theorem 3.18 *Suppose that the system (3.3.4) with (3.3.5) satisfies the following conditions:*

$$r_1^l > p_1^u + \lambda_1^u + d^u \xi_0^u,\ r_2^l > p_2^u + \lambda_2^u,\ b_2^l > p_1^u + p_2^u,\ b_{21}^l b_1^l > b_{11}^u (b_2^u - p_1^l - p_2^l),$$

$$\textit{and}\ \max\left\{\frac{r_1^u - p_1^l - \lambda_1^l - d^l \xi_0^l}{\beta_1^l}, \frac{r_2^u - p_2^l - \lambda_2^l}{\beta_2^l}, \frac{b_{21}^u b_1^u - b_{11}^l (b_2^l - p_1^u - p_2^u)}{b_{11}^l b_{22}^l}\right\} < \Delta_4$$

$$< \min\left\{\frac{b_1^l}{b_{12}^u}, \frac{b_{21}^l b_1^l - b_{11}^u (b_2^u - p_1^l - p_2^l)}{b_{21}^l b_{12}^u}\right\}.$$

Then, the system (3.3.4) with initial conditions (3.3.5) is uniformly persistent.

Proof Let $(X(t), Y_1(t), Y_2(t), Y_2(t))^{\mathrm{T}}$ be any solution of (3.3.4) with (3.3.5).

$$\therefore \dot{X}(t) \geq X(t)[b_1^l - b_{11}^u X(t) - b_{12}^u \Delta_4],\ \forall\, t \geq T_3$$

(applying Theorem 3.17 with T_3 as defined there).

Now, $\Delta_4 < \dfrac{b_{21}^l b_1^l - b_{11}^u (b_2^u - p_1^l - p_2^l)}{b_{21}^l b_{12}^u}$, and $b_2^l > \theta_1^u + \theta_2^u$ together implies

$0 < \dfrac{b_2^u - p_1^l - p_2^l}{b_{21}^l} < \dfrac{b_1^l - b_{12}^u \Delta_4}{b_{11}^u}$. Let us choose σ_1 in such a way that

$$0 < \frac{b_2^u - p_1^l - p_2^l}{b_{21}^l} < \sigma_1 < \frac{b_1^l - b_{12}^u \Delta_4}{b_{11}^u} \Rightarrow b_{21}^l \sigma_1 > b_2^u - p_1^l - p_2^l. \tag{3.3.9}$$

$$\text{If } X(T_3) \ge \sigma_1, \text{ then } X(t) \ge \sigma_1, \ \forall\, t \ge T_3. \text{ If } X(T_3) < \sigma_1, \text{ and}$$

$$\text{let } \alpha_1 = X(T_3)(b_1^l - b_{11}^u\sigma_1 - b_{12}^u\Delta_4) > 0, \text{ then } \exists\, \epsilon_1 > 0, \text{ s.t.}$$

$$X(t) < \sigma_1, \text{ and } \dot{X}_1(t) > \alpha_1 > 0, \ \forall\, t \in [T_3, T_3 + \epsilon_1).$$

$$\text{Therefore, } \exists\, T_4 > T_3 > 0, \text{ s.t. } X(t) \ge \sigma_1, \ \forall\, t \ge T_4.$$

Define $\Omega(t) = \min\{Y_1(t), Y_2(t), Y_3(t)\}$. Then, calculating the lower-right derivative of $\Omega(t)$ along the solution of the system (3.3.4) with (3.3.5), we get:

$$\begin{aligned}
&(1) \text{ If } Y_1(t) \le Y_2(t) \le Y_3(t), \text{ or } Y_1(t) \le Y_3(t) \le Y_2(t), \text{ then } D_+\Omega(t) = \dot{Y}_1(t)\\
&\ge r_1^l Y_1(t) - p_1^u Y_1(t) - \lambda_1^u Y_1(t) - \mathrm{d}^u \xi_0^u Y_1(t) - \beta_1^u Y_1^2(t)\\
&= (r_1^l - p_1^u - \lambda_1^u - \mathrm{d}^u\xi_0^u) Y_1(t) - \beta_1^u Y_1^2(t), \ \forall\, t \ge 0.
\end{aligned} \tag{3.3.10}$$

$$\begin{aligned}
&(2) \text{ If } Y_2(t) \le Y_1(t) \le Y_3(t), \text{ or } Y_2(t) \le Y_3(t) \le Y_1(t), \text{ then } D_+\Omega(t) = \dot{Y}_2(t)\\
&\ge r_2^l Y_2(t) - p_2^u Y_2(t) - \lambda_2^u Y_2(t) - \beta_2^u Y_2^2(t)\\
&= (r_2^l - p_2^u - \lambda_2^u) Y_2(t) - \beta_2^u Y_2^2(t), \ \forall\, t \ge 0.
\end{aligned} \tag{3.3.11}$$

$$\begin{aligned}
&(3) \text{ If } Y_3(t) \le Y_1(t) \le Y_2(t), \text{ or } Y_3(t) \le Y_2(t) \le Y_1(t), \text{ then } D_+\Omega(t) = \dot{Y}_3(t)\\
&\ge -b_2^u Y_3(t) - b_{22}^u Y_3^2(t) + b_{21}^l \sigma_1 Y_3(t-\tau) + (p_1^l + p_2^l) Y_3(t)\\
&= -(b_2^u - p_1^l - p_2^l) Y_3(t) - b_{22}^u Y_3^2(t) + b_{21}^l \sigma_1 Y_3(t-\tau), \ \forall\, t \ge T_4 + \tau.
\end{aligned} \tag{3.3.12}$$

Let us choose σ_2 in such a way that $0 < \sigma_2 < \min\left\{\frac{r_1^l - p_1^u - \lambda_1^u - d^u\xi_0^u}{\beta_1^u}, \frac{r_2^l - p_2^u - \lambda_2^u}{\beta_2^u}\right\}$.

From (3.3.10) and (3.3.11), we can get the following:

(i) If $\Omega(0) = \min\{Y_1(0), Y_2(0), Y_3(0)\} \ge \sigma_2$, then $\min\{Y_1(t), Y_2(t), Y_3(t)\} \ge \sigma_2$, $\forall\, t \ge 0$.
(ii) If $\Omega(0) = \min\{Y_1(0), Y_2(0), Y_3(0)\} < \sigma_2$, and let $\alpha_2 = \min[Y_1(0)\{r_1^l - p_1^u - \lambda_1^u - d^u\xi_0^u - \beta_1^u\sigma_2\}, Y_2(0)\{r_2^l - p_2^u - \lambda_2^u - \beta_2^u\sigma_2\}] > 0$, we consider the following two cases:

$$\text{(a) } \Omega(0) = Y_1(0) < \sigma_2,$$

$$\text{(b) } \Omega(0) = Y_2(0) < \sigma_2.$$

If (a) holds, then $\exists\, \epsilon_2 > 0$, s.t. if $t \in [0, \epsilon_2)$, $\Omega(t) = Y_1(t) < \sigma_2$, and we have $D_+\Omega(t) = \dot{Y}_1(t) > \alpha_2 > 0$.
If (b) holds, then $\exists\, \epsilon_3 > 0$, s.t. if $t \in [0, \epsilon_3)$, $\Omega(t) = Y_2(t) < \sigma_2$, and we have $D_+\Omega(t) = \dot{Y}_2(t) > \alpha_2 > 0$.

If (3.3.12) holds, then by (3.3.9) and Theorem 3.16, $\exists\, T_5 \geq T_4 + \tau > 0$, s.t.

$$Y_3(t) > \sigma_3, \ \forall\, t \geq T_5 \geq T_4 + \tau > 0, \ \text{where}\ \sigma_3 < \frac{b_{21}^l \sigma_1 - (b_2^u - p_1^l - p_2^l)}{b_{22}^u}.$$

From the discussions, we can conclude that $\exists\, T_6 \geq T_5 \geq T_4 + \tau > T_3 > 0$, s.t. $X(t) \geq \sigma_1$ and $Y_i(t) \geq \sigma_4$ $(i = 1, 2, 3)$, $\forall\, t \geq T_6$, where $0 < \frac{b_2^u - p_1^l - p_2^l}{b_{21}^l} < \sigma_1 < \frac{b_1^l - b_{12}^u \Delta_4}{b_{11}^u}$, and $0 < \sigma_4 < \min\left\{\frac{r_1^l - p_1^u - \lambda_1^u - d^u \xi_0^u}{\beta_1^u}, \frac{r_2^l - p_2^u - \lambda_2^u}{\beta_2^u}, \frac{b_{21}^l \sigma_1 - (b_2^u - p_1^l - p_2^l)}{b_{22}^u}\right\}$.

From the discussions, we can conclude that $\exists\, T_6 \geq T_5 \geq T_4 + \tau > T_3 > 0$ s.t. every solution of system (3.3.4) with (3.3.5) eventually enters and remains in the region $B = \{(X, Y_1, Y_2, Y_3) \mid \sigma \leq X, Y_i \leq \Delta,\ i = 1, 2, 3\}$, $\forall\, t \geq T_6$, where $\sigma = \min\{\sigma_1, \sigma_4\}$ and $\Delta = \max\{\Delta_1, \Delta_4\}$, Δ_1 is defined in Theorem 3.17. Hence, the theorem is proved. □

The results of Theorem 3.18 express the fact that lower values $\lambda_1(t)$ (instantaneous per capita death rate function of the individuals in Υ_1), $\lambda_2(t)$ (instantaneous per capita death rate function of the individuals in Υ_2), $\xi_0(t)$ (concentration of toxicant in the immature predator organism in nonprotective patch (living in Υ_1) at time t), $\xi_e(t)$ (concentration of toxicant in the environment at time t) and $d(t)$ (dose response parameter of the individuals in Υ_1 to the organismal toxicant concentration at time t) lead to make the underlying system uniformly persistent.

3.3.3 Global Asymptotic Stability of Periodic Solution

From our day-to-day experience, it is known that the biological and environmental parameters are subjected to fluctuation in time. The effects of a periodically varying environment have a crucial selective forces on systems in a fluctuating (random) environment. For investigating this kind of natural phenomenon, in the model, the coefficients should be periodic functions of time. Let that system (3.3.4) is a periodic system with (3.3.5) satisfying all conditions of Theorem 3.18. The sufficient conditions for all solutions of system (3.3.4) with (3.3.5) to converge to a periodic solution are derived in Theorem 3.21.

Theorem 3.19 *(Brouwer) (Hirsch 1980). Assume that the continuous operator f maps the closed and bounded convex set $C \in R^n$ onto itself, then the operator f has at least one fixed point in set C.*

Theorem 3.20 *If the periodic system (3.3.4) with (3.3.5) satisfies the conditions of Theorem 3.18, then there exists at least one strictly positive periodic solution of this system.*

Proof From Theorems 2.5, 2.6 and 2 in Teng and Chen (1999), it is easy to derive at least one strictly positive periodic solution of the system (3.3.4) with (3.3.5). □

Theorem 3.21 *If the system (3.3.4) with (3.3.5) has a periodic solution, then it is globally asymptotically stable provided*

$$\liminf_{t\to\infty} L_i(t) > 0,\ (i=1,2),\ \lambda_1^l + d^l\xi_0^l + 2\sigma\beta_1^l + \delta_1^l - \delta_2^u > 0,\ \lambda_2^l + 2\sigma\beta_2^l + \delta_2^l - \delta_1^u > 0,$$

where $L_1(t) = b_{11}^l - \Delta b_{21}(t+\tau)$, $L_2(t) = b_2^l + 2\sigma b_{22}^l - b_{12}^u - r_1^u - r_2^u - \Delta b_{21}(t+\tau)$, σ, Δ *are defined as in Theorem 3.18.*

Proof Let $(v(t), \omega_1(t), \omega_2(t), \omega_3(t))^{\mathrm{T}}$ be a periodic solution of system (3.3.4) with (3.3.5). Let $(X(t), Y_1(t), Y_2(t), Y_3(t))^{\mathrm{T}}$ be any positive solution of (3.3.4) with (3.3.5). Suppose that

$$\Omega_1(t) = \mid \ln X(t) - \ln v(t) \mid + \sum_{i=1}^{3} \mid Y_i(t) - \omega_i(t) \mid . \tag{3.3.13}$$

Calculating the upper-right derivative of $\Omega_1(t)$ along the positive solution of the system (3.3.4) with (3.3.5), we get

$$\begin{aligned}
D^+\Omega_1(t) &= \left(\frac{\dot X(t)}{X(t)} - \frac{\dot v(t)}{v(t)}\right)\operatorname{sgn}(X(t) - v(t)) + \sum_{i=1}^{3}(\dot Y_i(t) - \dot\omega_i(t))\operatorname{sgn}(Y_i(t) - \omega_i(t)) \\
&= \operatorname{sgn}(X(t) - v(t))\{-b_{11}(t)(X(t) - v(t)) - b_{12}(t)(Y_3(t) - \omega_3(t))\} \\
&\quad + \operatorname{sgn}(Y_1(t) - \omega_1(t))\{r_1(t)(Y_3(t) - w_3(t)) - (p_1(t) + \lambda_1(t) + d(t)\xi_0(t))(Y_1(t) \\
&\quad - \omega_1(t)) - \beta_1(t)(Y_1^2(t) - \omega_1^2(t)) + \delta_1(t)((Y_2(t) - \omega_2(t)) - (Y_1(t) - \omega_1(t)))\} \\
&\quad + \operatorname{sgn}(Y_2(t) - \omega_2(t))\{r_2(t)(Y_3(t) - \omega_3(t)) - (p_2(t) + \lambda_2(t))(Y_2(t) \\
&\quad - \omega_2(t)) - \beta_2(t)(Y_2^2(t) - \omega_2^2(t)) + \delta_2(t)((Y_1(t) - \omega_1(t)) - (Y_2(t) - \omega_2(t)))\} \\
&\quad + \operatorname{sgn}(Y_3(t) - \omega_3(t))\{-b_2(t)(Y_3(t) - \omega_3(t)) - b_{22}(t)(Y_3^2(t) - \omega_3^2(t)) + b_{21}(t)Y_3(t-\tau)(X(t-\tau) - v(t-\tau)) \\
&\quad + b_{21}(t)v(t-\tau)(Y_3(t-\tau) - \omega_3(t-\tau)) + p_1(t)(Y_1(t) - \omega_1(t)) + p_2(t)(Y_2(t) - \omega_2(t))\} \\
&\le -b_{11}^l \mid X(t) - v(t) \mid -(\lambda_1^l + \mathrm{d}^l\xi_0^l + 2\sigma\beta_1^l + \delta_1^l - \delta_2^u) \mid Y_1(t) - \omega_1(t) \mid
\end{aligned}$$

$$-(\lambda_2^l + 2\sigma\beta_2^l + \delta_2^l - \delta_1^u) \mid Y_2(t) - \omega_2(t) \mid -(b_2^l + 2\sigma b_{22}^l - b_{12}^u - r_1^u - r_2^u) \mid Y_3(t) - \omega_3(t) \mid$$

$$+ b_{21}(t)Y_3(t-\tau) \mid X(t-\tau) - v(t-\tau) \mid + b_{21}(t)v(t-\tau) \mid Y_3(t-\tau) - \omega_3(t-\tau) \mid . \quad (3.3.14)$$

$$\text{Let, } \Omega_2(t) = \int_{t-\tau}^{t} b_{21}(s+\tau)Y_3(s) \mid X(s) - v(s) \mid \mathrm{d}s + \int_{t-\tau}^{t} b_{21}(s+\tau)v(s) \mid Y_3(s) - \omega_3(s) \mid \mathrm{d}s. \quad (3.3.15)$$

Then,

$$D^+\Omega_2(t) = b_{21}(t+\tau)Y_3(t) \mid X(t) - v(t) \mid - b_{21}(t)Y_3(t-\tau) \mid X(t-\tau) - v(t-\tau) \mid$$

$$+ b_{21}(t+\tau)v(t) \mid Y_3(t) - \omega_3(t) \mid - b_{21}(t)v(t-\tau) \mid Y_3(t-\tau) - \omega_3(t-\tau) \mid . \quad (3.3.16)$$

$$\text{Let, } \Omega(t) = \Omega_1(t) + \Omega_2(t). \quad (3.3.17)$$

Then,

$$D^+\Omega(t) \le -(b_{11}^l - \Delta b_{21}(t+\tau)) \mid X(t) - v(t) \mid -(\lambda_1^l + \mathrm{d}^l\xi_0^l + 2\sigma\beta_1^l + \delta_1^l - \delta_2^u) \mid Y_1(t) - \omega_1(t) \mid$$

$$-(\lambda_2^l + 2\sigma\beta_2^l + \delta_2^l - \delta_1^u) \mid Y_2(t) - \omega_2(t) \mid -(b_2^l + 2\sigma b_{22}^l - b_{12}^u - r_1^u - r_2^u - \Delta b_{21}(t+\tau)) \mid Y_3(t) - \omega_3(t) \mid$$

$$= -L_1(t) \mid X(t) - v(t) \mid -(\lambda_1^l + \mathrm{d}^l\xi_0^l + 2\sigma\beta_1^l + \delta_1^l - \delta_2^u) \mid Y_1(t) - \omega_1(t) \mid$$

$$-(\lambda_2^l + 2\sigma\beta_2^l + \delta_2^l - \delta_1^u) \mid Y_2(t) - \omega_2(t) \mid - L_2(t) \mid Y_3(t) - \omega_3(t) \mid, \quad (3.3.18)$$

where $L_1(t) = b_{11}^l - \Delta b_{21}(t+\tau)$, $L_2(t) = b_2^l + 2\sigma b_{22}^l - b_{12}^u - r_1^u - r_2^u - \Delta b_{21}(t+\tau)$.

By assumption: $\lambda_1^l + d^l\xi_0^l + 2\sigma\beta_1^l + \delta_1^l - \delta_2^u > 0$, $\lambda_2^l + 2\sigma\beta_2^l + \delta_2^l - \delta_1^u > 0$, and $\exists\ \mu_1 > 0,\ \mu_2 > 0, T^* \ge T_6 > 0$, ($T_6$ is defined as in Theorem 3.18), such that $L_1(t) \ge \mu_1,\ L_2(t) \ge \mu_2,\ \forall\ t \ge T^*$. Integrating both sides of (3.3.18) in the interval $[T^*, t]$:

$$\Omega(t) + \int_{T^*}^{t} L_1(s) \mid X(s) - v(s) \mid ds + (\lambda_1^l + \mathrm{d}^l\xi_0^l + 2\sigma\beta_1^l + \delta_1^l - \delta_2^u) \int_{T^*}^{t} \mid Y_1(s) - \omega_1(s) \mid \mathrm{d}s$$

$$+(\lambda_2^l + 2\sigma\beta_2^l + \delta_2^l - \delta_1^u) \int_{T^*}^{t} \mid Y_2(s) - \omega_2(s) \mid \mathrm{d}s + \int_{T^*}^{t} L_2(s) \mid Y_3(s) - \omega_3(s) \mid \mathrm{d}s \le \Omega(T^*)$$

$$\Rightarrow \Omega(t) + \mu_1 \int_{T^*}^{t} \mid X(s) - v(s) \mid \mathrm{d}s + (\lambda_1^l + \mathrm{d}^l \xi_0^l + 2\sigma\beta_1^l + \delta_1^l - \delta_2^u) \int_{T^*}^{t} \mid Y_1(s) - \omega_1(s) \mid \mathrm{d}s$$

$$+(\lambda_2^l + 2\sigma\beta_2^l + \delta_2^l - \delta_1^u) \int_{T^*}^{t} \mid Y_2(s) - \omega_2(s) \mid \mathrm{d}s + \mu_2 \int_{T^*}^{t} \mid Y_3(s) - \omega_3(s) \mid \mathrm{d}s \leq \Omega(T^*)$$

$$\Rightarrow \Omega(t) \text{ is bounded on } [T^*, \infty) \text{ and also } \int_{T^*}^{\infty} \mid X(s) - v(s) \mid \mathrm{d}s < \infty,$$

$$\int_{T^*}^{\infty} \mid Y_i(s) - \omega_i(s) \mid \mathrm{d}s < \infty, \ (i = 1, 2, 3).$$

By Theorem 3.17, $\mid X(s) - v(s) \mid, \ \mid Y_i(s) - \omega_i(s) \mid \ (i = 1, 2, 3)$ are bounded on $[T^*, \infty)$. On the other side, it is easy to observe that $\dot{X}(t), \dot{v}(t), \dot{Y}_i(t)$ and $\dot{\omega}_i(t), \ (i = 1, 2, 3)$ are bounded for $t \geq T^*$. Therefore, $\mid X(t) - v(t) \mid, \ \mid Y_i(t) - \omega_i(t) \mid \ (i = 1, 2, 3)$ are uniformly continuous in the interval $[T^*, \infty)$. Applying Barbalat's Lemma (Gopalsamy 1992):

$$\mid X(t) - v(t) \mid \rightarrow 0, \ \mid Y_i(t) - \omega_i(t) \mid \rightarrow 0 \text{ as } t \rightarrow \infty, \ i = 1, 2, 3.$$

Hence, the theorem is proved. □

Remark: The diffusion coefficient functions and time delay have no effect on the uniformly persistent of the system (3.3.4) with (3.3.5), but they have some effect on the global asymptotic stability of this system.

3.3.4 Discussion

In this article, we have analysed the effects of pollution in a nonautonomous Lotka–Volterra type prey–predator system together with diffusional migration among the immature predator population between protective and nonprotective patches. To take into account the gestation of the predator a constant delay among the matured predator has been incorporated. It is also assumed that the mature adult predator can only take part to the reproduction of the predator biomass and that the change rate of predator depends on the densities of the prey and of the predator presented at some previous time. A stage structure for the predator has been incorporated. It is assumed that the toxicants have no effect on the mature species.

The toxic of polluted population comes from the environment where they live. Suppose that the environment of patch Υ_1 is polluted nonprotective region, and patch Υ_2 is the protective region for the immature predator population. To protect the existing polluted population, artificial method can be employed: using own purification function of the population, toxin in the body of the individuals in nonprotective patch can be removed and then bringing them into protective patch. The scale or degree of the protective zone can be controlled through diffusive coefficients $\delta_i(t),\ (i = 1, 2)$.

Here we have obtained the survival threshold of each population of the system under consideration. Our analysis reveals the fact that lower values of $\lambda_1(t)$ (instantaneous per capita death rate function of nonprotective immature predator population), $\lambda_2(t)$ (instantaneous per capita death rate function of protective immature predator population), $\xi_0(t)$ (concentration of toxicant in the immature predator organism in nonprotective patch at time t), $\xi_e(t)$ (concentration of toxicant in the environment at time t) and $d(t)$ (dose response parameter of nonprotective immature predator species to the organismal toxicant concentration at time t) lead to make the system uniformly persistent. By constructing an appropriate function (called Lyapunov function), we have got sufficient conditions to confirm that if this system admits a positive periodic solution, then it is globally asymptotically stable. We have seen that the diffusion coefficient functions and time delay have no effect on the uniformly persistent of the system (3.3.4) with (3.3.5) but they have some effect on the global asymptotic stability of this system.

References

Gopalsamy, K.: Stability and Oscillation in Delay Differential Equations of Population Dynamics. Kluwer Academic Press, Netherlands Dordrecht (1992)

Hale, J.K.: Theory of Functional Differential Equations. Springer-Verlag, New York (1977)

Hirsch, M.W.: Differential Topology. Springer, New York (1980)

Manning, W.J., Feder, W.A.: Biomonitoring Air Pollutants with Plants. Applied Science Publishers, Barking (1980)

Song, X.Y., Chen, L.S.: Optimal harvesting and stability with stage-structure for a two species competitive system. Math. Biosci. **170**, 173–186 (2001)

Teng, Z., Chen, L.: The positive periodic solutions of periodic Kolmogorov type systems with delays. Acta Math. Appl. Sin. **22**, 446–456 (1999). (in Chinese)

Xiao, Y., Chen, L.: Effects of toxicants on a stage-structured population growth model. Appl. Math. Comput. **123**, 63–73 (2001)

Zhou, X., Shi, X., Song, X.: Analysis of nonautonomous predator-prey model with nonlinear diffusion and time delay. Appl. Math. Comput. **196**, 129–136 (2008)

Chapter 4
Dynamical Models of Single-Species System Under the Influence of Environmental Noise

4.1 Introduction

This chapter aims to study the influence of environmental noise in Gompertzian and Logistic growth models. The stability behaviour of these models in a deterministic environment and the corresponding model in a stochastic environment has been analysed here.

4.2 Influence of Environmental Noise in Gompertzian Growth Model

In this article, we have studied the influence of environmental noise in Gompertzian growth model. The relationship between stability behaviour of this model in a deterministic environment and the corresponding model in a stochastic environment has also been discussed here.

4.2.1 Gompertzian Growth with Random Birth Rate

Let us consider the following Gompertzian growth model:

$$\frac{\mathrm{d}N}{\mathrm{d}t} = \{m - a \ln N\} N, \tag{4.2.1}$$

where N can be either population of a species or size of an organ or tumor, m is the net birth rate and $a(> 0)$ measures the intraspecific competition among the species which is assumed to be constant. This is the most generally accepted model of tumor growth.

© Springer Nature Singapore Pte Ltd. 2021

G. Samanta, *Deterministic, Stochastic and Thermodynamic Modelling of some Interacting Species*, Forum for Interdisciplinary Mathematics,
https://doi.org/10.1007/978-981-16-6312-3_4

To take into account of the random environment, we extend model (4.2.1) to the form of the stochastic differential equation:

$$\frac{\mathrm{d}N}{\mathrm{d}t} = \{m_0 - a\ln N\}N + Nu, \tag{4.2.2}$$

where the environmental parameter m varies randomly and

$$m = m_0 + u(t), \tag{4.2.3}$$

$m_0(> 0)$ is a constant, being the mean value of m, and $u(t)$ is a *colored noise* or *Ornstein–Uhlenbeck process* which is more realistic noise than *white noise*. The mathematical expectation and correlation function of the process $u(t)$ are given by

$$\langle u(t)\rangle = 0 \text{ and } \langle u(t_1)u(t_2)\rangle = \epsilon\delta_0 \exp\{-\delta_0|t_1 - t_2|\}, \ (\epsilon, \delta_0 > 0) \tag{4.2.4}$$

where δ_0^{-1}, ϵ are given constants describing the correlation time and the intensity of the noise, respectively, and this is the solution of the following stochastic differential equation (Uhlenbeck and Ornstein 1954):

$$\frac{\mathrm{d}u}{\mathrm{d}t} = -\delta_0 u + \delta_0\sqrt{2\epsilon}\,\gamma \tag{4.2.5}$$

where $\gamma(t)$ denotes the standard zero mean Gaussian white noise.

From (4.2.2), we have

$$\frac{\mathrm{d}x}{\mathrm{d}t} = m_0 - ax + u, \ (x = \ln N). \tag{4.2.6}$$

From (4.2.5) and (4.2.6), we obtain the following stochastic differential equation:

$$\frac{\mathrm{d}^2x}{\mathrm{d}t^2} + h\left(x, \frac{\mathrm{d}x}{\mathrm{d}t}\right) = \delta_0\sqrt{2\epsilon}\,\gamma \tag{4.2.7}$$

where $\gamma(t)$ is a standard zero-mean Gaussian white noise characterized by

$$\langle\gamma(t)\rangle = 0, \ \langle\gamma(t_1)\gamma(t_2)\rangle = \delta(t_1 - t_2) \tag{4.2.8}$$

and

$$h\left(x, \frac{\mathrm{d}x}{\mathrm{d}t}\right) = (\delta_0 + a)\frac{\mathrm{d}x}{\mathrm{d}t} - \delta_0(m_0 - ax). \tag{4.2.9}$$

4.2.2 Stationary Probability Density

The reduced Fokker–Planck equation is given by Cai and Lin (1988):

$$x_2 \frac{\partial p_s}{\partial x_1} - \frac{\partial}{\partial x_2} h(x_1, x_2) p_s - \delta_0^2 \epsilon \frac{\partial^2 p_s}{\partial x_2^2} = 0, \text{ where } x_1 = x(t), x_2 = \frac{dx}{dt} \tag{4.2.10}$$

and $p_s \equiv p_s(x_1, x_2)$ is the stationary probability density.

After some simplifications using the results of Cai and Lin (1988), we have

$$p_s(x_1, x_2) = C \exp\left\{-\frac{(\delta_0 + a)}{2\delta_0^2 \epsilon}(x_2^2 - 2m_0\delta_0 x_1 + a\delta_0 x_1^2)\right\} \tag{4.2.11}$$

where C is a normalization constant.

$$\therefore p_s(x_1) = \frac{1}{\sigma\sqrt{2\pi}} \exp\left\{-\frac{\left(x_1 - \frac{m_0}{a}\right)^2}{2\sigma^2}\right\}, \text{ where } \sigma^2 = \frac{\delta_0 \epsilon}{(\delta_0 + a)a}. \tag{4.2.12}$$

That is, $x = \ln N$ is normally distributed with mean m_0/a and variance $\sigma^2 = \frac{\delta_0 \epsilon}{(\delta_0 + a)a}$.

Therefore, the stationary probability density of N is given by

$$p_s(n) = \frac{1}{n\sigma\sqrt{2\pi}} \exp\left\{-\frac{\left(\ln n - \frac{m_0}{a}\right)^2}{2\sigma^2}\right\} \Rightarrow \langle N \rangle = \exp\left\{\frac{m_0}{a} + \frac{\sigma^2}{2}\right\}. \tag{4.2.13}$$

The root-mean-square relative fluctuation of the population about this mean can be shown to be

$$\frac{\sqrt{\langle (N - \langle N \rangle)^2 \rangle}}{\langle N \rangle} = \sqrt{(e^{\sigma^2} - 1)}, \text{ where } \sigma^2 = \frac{\delta_0 \epsilon}{(\delta_0 + a)a}. \tag{4.2.14}$$

Now, in the deterministic environment, m is constant and the eigen value of the 1×1 interaction matrix is $-a$, which is negative, showing that the equilibrium point is stable. In the stochastic environment, whose random fluctuation has intensity ϵ, the stability provided by the population interaction dynamics is again characterized by a. It is no longer enough that $a > 0$, for if $\epsilon \gg a$, population exhibits large fluctuation, which rapidly lead to extinction. For $a > \epsilon$, but not much greater, population are likely to undergo significant fluctuations, even though they persist for long times. Finally, if $a \gg \epsilon$, population fluctuations are relatively small, and the environment is effectively deterministic. It is also noted that as $\delta_0 \to \infty$, the colored noise $u(t)$

becomes the white noise, and we get the same results for white noise idealization (in this case, Stratonovich calculus is used). If we take the Itô interpretation of stochastic differential equation for white noise idealization, then we also get the same results.

4.2.3 *Gompertzian Growth with Random Birth Rate and Crowding Coefficient*

In this section, we are considering the following form of the Gompertzian growth model:

$$\frac{dN}{dt} = \{m(t) - a(t)\ln N\}N, \; N(0) = 1, \tag{4.2.15}$$

where

$$m(t) = m_0(1 + \epsilon\cos pt) + u_1(t), \; a(t) = a_0(1 + \epsilon\cos pt) + u_2(t). \tag{4.2.16}$$

Here, $u_1(t)$ is a real Gaussian environmental noise with mean zero and $u_2(t)$ is a *colored noise* or *Ornstein–Uhlenbeck process* which is more realistic noise than *white noise*, satisfying the conditions:

$$\langle u_2(t)\rangle = 0 \text{ and } \langle u_2(t_1)u_2(t_2)\rangle = \epsilon_2\delta_2\exp\{-\delta_2|t_1 - t_2|\}, \; (\epsilon_2, \delta_2 > 0), \tag{4.2.17}$$

where $\langle\cdot\rangle$ represents the average over the ensemble of the stochastic process. Here, mean of $m(t)$ and $a(t)$ be periodic functions of time with period $\tau = \frac{2\pi}{p}$, m_0, a_0 are positive constants and $0 < \epsilon \ll 1$. Such periodic nature of the growth and interaction rates may arise out of the seasonal changes in the environment or from periodic nature of mating of animals and subsequently the breeding of these species. Also, such fluctuations in the interaction terms occur due to periodic variations of common resources such as food. Here, it is assumed that $u_1(t)$ and $u_2(t)$ are independent.

From (4.2.15), we have

$$\frac{dx}{dt} + a(t)x = m(t), \; x(0) = 0, \; (x = \ln N) \Rightarrow x(t) = \int_0^t m(u)\exp\left\{-\int_u^t a(s)ds\right\}du. \tag{4.2.18}$$

For calculating the moments of $x(t)$, a significant simplification results if we introduce the process:

$$Z(t)=\int_0^t \exp\left\{-\alpha\, m(u)-\int_u^t a(s)\mathrm{d}s\right\}\mathrm{d}u \Rightarrow x(t)=-\left[\frac{\partial}{\partial\alpha}Z(t)\right]_{\alpha=0}. \quad (4.2.19)$$

$$\therefore \langle x(t)\rangle=-\left[\frac{\partial}{\partial\alpha}\langle Z(t)\rangle\right]_{\alpha=0},\ \text{where } \langle Z(t)\rangle=\int_0^t\left\langle \exp\left\{-\alpha\, m(u)-\int_u^t a(s)\mathrm{d}s\right\}\right\rangle \mathrm{d}u. \quad (4.2.20)$$

Let

$$U(u)=\int_u^t a(s)\mathrm{d}s \quad (4.2.21)$$

and let $\phi(u_1, u_2)$ denotes the joint characteristics function of the two Gaussian random variables $m(u)$ and $U(u)$, then

$$\left\langle \exp\left\{-\alpha\, m(u)-\int_u^t a(s)\mathrm{d}s\right\}\right\rangle=\phi(i\alpha,\alpha)$$
$$=\exp\left[-\{\alpha\langle m(u)\rangle+\langle U(u)\rangle\}+\frac{1}{2}\{\alpha^2\sigma_m^2+\sigma_u^2\}\right]. \quad (4.2.22)$$

$$\therefore \langle x(t)\rangle=-\left[\frac{\partial}{\partial\alpha}\langle Z(t)\rangle\right]_{\alpha=0}=\int_0^t\langle m(u)\rangle \exp\left\{-\langle U(u)\rangle+\frac{1}{2}\sigma_u^2\right\}\mathrm{d}u, \quad (4.2.23)$$

where

$$\langle m(u)\rangle=m_0(1+\epsilon\cos pu),\ \langle U(u)\rangle=\int_u^t\langle a(s)\rangle\mathrm{d}s$$
$$=\int_u^t a_0(1+\epsilon\cos ps)\mathrm{d}s=a_0(t-u)+\frac{\epsilon}{p}(\sin pt-\sin pu) \quad (4.2.24)$$

and

$$\sigma_u^2(t,u) = \langle (U(u) - \langle U(u)\rangle)^2 \rangle = \int_u^t \int_u^t \langle (a(s_1) - \langle a(s_1)\rangle)\,(a(s_2) - \langle a(s_2)\rangle)\rangle \mathrm{d}s_1 \mathrm{d}s_2$$
$$= \epsilon_2\delta_2 \int_u^t \int_u^t \exp\{-\delta_2|s_2 - s_1|\}\,\mathrm{d}s_1\mathrm{d}s_2 = 2\epsilon_2\left[t - u + \frac{1}{\delta_2}\left\{\mathrm{e}^{\delta_2(u-t)} - 1\right\}\right] \tag{4.2.25}$$

$$\therefore \langle x(t)\rangle = m_0 \int_0^t \exp\left[a_0(u-t) + \frac{\epsilon}{p}(\sin pu - \sin pt) + \epsilon_2\left\{t - u + \frac{1}{\delta_2}(\mathrm{e}^{\delta_2(u-t)} - 1)\right\}\right]$$
$$\times (1 + \epsilon\cos pu)\mathrm{d}u = m_0 \exp\left\{(\epsilon_2 - a_0)t - \frac{\epsilon}{p}\sin pt\right\}$$
$$\times \int_0^t (1 + \epsilon\cos pu)\exp\left\{-(\epsilon_2 - a_0)u + \frac{\epsilon}{p}\sin pu + \frac{\epsilon_2}{\delta_2}(\mathrm{e}^{\delta_2(u-t)} - 1)\right\} du \tag{4.2.26}$$

Now, for $\epsilon_2 > a_0$:

$$\int_0^t (1 + \epsilon\cos pu)\exp\left\{-(\epsilon_2 - a_0)u + \frac{\epsilon}{p}\sin pu + \frac{\epsilon_2}{\delta_2}(\mathrm{e}^{\delta_2(u-t)} - 1)\right\} du$$
$$> \exp\left\{-\left(\frac{\epsilon}{p} + \frac{\epsilon_2}{\delta_2}\right)\right\} \int_0^t (1 + \epsilon\cos pu)\exp\{-(\epsilon_2 - a_0)u\}\,du$$
$$= \exp\left\{-\left(\frac{\epsilon}{p} + \frac{\epsilon_2}{\delta_2}\right)\right\}\left[\left\{\frac{1}{\epsilon_2 - a_0} + \frac{\epsilon(\epsilon_2 - a_0)}{(\epsilon_2 - a_0)^2 + p^2}\right\}\right.$$
$$+\left\{\frac{\epsilon}{(\epsilon_2 - a_0)^2 + p^2}(p\sin pt - (\epsilon_2 - a_0)\cos pt)\right.$$
$$\left.\left. -\frac{1}{(\epsilon_2 - a_0)}\right\}\exp\{-(\epsilon_2 - a_0)t\}\right]. \tag{4.2.27}$$

$$\therefore \langle x(t)\rangle > m_0 \exp\left\{-\left(\frac{\epsilon}{p}(1 + \sin pt) + \frac{\epsilon_2}{\delta_2}\right)\right\}$$
$$\times\left[\left\{\frac{\epsilon}{(\epsilon_2 - a_0)^2 + p^2}(p\sin pt - (\epsilon_2 - a_0)\cos pt)\right\} - \frac{1}{(\epsilon_2 - a_0)}\right.$$
$$\left. + \left\{\frac{1}{\epsilon_2 - a_0} + \frac{\epsilon(\epsilon_2 - a_0)}{(\epsilon_2 - a_0)^2 + p^2}\right\}\exp\{(\epsilon_2 - a_0)t\}\right], \tag{4.2.28}$$

showing that the instability is a consequence of stochastic parametric perturbation for $\epsilon_2 > a_0$, whereas the deterministic case is always stable (for $m_0, a_0 > 0$, the usual assumption).

For $\epsilon_2 = a_0$:

$$\begin{aligned}
\therefore \langle x(t)\rangle &= m_0 \exp\left\{-\frac{\epsilon}{p}\sin pt\right\}\int_0^t (1+\epsilon\cos pu)\exp\left\{\frac{\epsilon}{p}\sin pu + \frac{\epsilon_2}{\delta_2}(e^{\delta_2(u-t)}-1)\right\}du \\
&> m_0 \exp\left\{-\frac{\epsilon}{p}\sin pt\right\}\int_0^t (1+\epsilon\cos pu)\exp\left\{-\frac{\epsilon}{p}\sin pu - \frac{\epsilon_2}{\delta_2}\right\}du \\
&= m_0\left\{t+\frac{\epsilon}{p}\sin pt\right\}\exp\left\{-\left(\frac{\epsilon}{p}(1+\sin pt)+\frac{\epsilon_2}{\delta_2}\right)\right\}.
\end{aligned} \tag{4.2.29}$$

Therefore, the system is unstable due to the influence of environmental noises on the growth parameters for $\epsilon_2 = a_0$.

For $\epsilon_2 < a_0$:

$$\begin{aligned}
\therefore \langle x(t)\rangle &< m_0 \exp\left\{\frac{\epsilon}{p}(1-\sin pt)+\frac{\epsilon_2}{\delta_2}\right\} \\
&\times\left[\left\{\frac{\epsilon}{(a_0-\epsilon_2)^2+p^2}(p\sin pt+(a_0-\epsilon_2)\cos pt)\right\}+\frac{1}{(a_0-\epsilon_2)}\right. \\
&\left.-\left\{\frac{1}{a_0-\epsilon_2}+\frac{\epsilon(a_0-\epsilon_2)}{(a_0-\epsilon_2)^2+p^2}\right\}\exp\{-(a_0-\epsilon_2)t\}\right].
\end{aligned} \tag{4.2.30}$$

This indicates that, for $\epsilon_2 < a_0$, the system remains stable under the influence of environmental noises.

It is also noted that as $\delta_2 \to \infty$ the colored noise $u_2(t)$ becomes the white noise and we get the same results for white noise idealization (in this case, Stratonovich calculus is used).

4.3 On Stability and Fluctuation in Logistic Growth Model in a Random Environment

This article aims to study the influence of colored noise on logistic growth model. The relationship between stability behaviour of this model in a deterministic environment and the corresponding model in a stochastic environment has also been discussed here. The discussion is based on the complex stochastic averaging technique.

4.3.1 Logistic Growth: Stochastic Differential Equation

The Verhulst (or logistic) equation of growth of a single species population is given by

$$\frac{dN}{dt} = aN - bN^2 = aN\left(1 - \frac{N}{K}\right), \tag{4.3.1}$$

where $N(t)$ is the population size at any time t, $a(> 0)$ represents the intrinsic growth rate and $K = \frac{a}{b}(> 0)$ is the environmental carrying capacity. Here, a, b are assumed to be constants. This type of equation occurs in different cases of biophysical system. The stochastic extension of Eq. (4.3.1) is given by

$$\frac{dN}{dt} = aN - bN^2 + u(t), \tag{4.3.2}$$

where $u(t)$ is the random perturbation term which is due to the overall effect of the numerous microscopic, unknown or partially known internal fluctuation of the state variable N. We have called the randomness of perturbation to be internal in the sense that it is due to the internal effect of mutual interactions of the individuals of the same species without any consideration of the randomness of the environmental parameters. The random perturbation term $u(t)$ is assumed to be a *colored noise* or *Ornstein–Uhlenbeck process* which is more realistic noise than *white noise*. The mathematical expectation and correlation function of the process $u(t)$ are given by

$$\langle u(t)\rangle = 0 \text{ and } \langle u(t_1)u(t_2)\rangle = \epsilon\delta_0 \exp\{-\delta_0|t_1 - t_2|\},\ (\epsilon, \delta_0 > 0), \tag{4.3.3}$$

where δ_0^{-1}, ϵ are given constants describing the correlation time and the intensity of the noise, respectively, and this is the solution of the following stochastic differential equation (Uhlenbeck and Ornstein 1954):

$$\frac{du}{dt} = -\delta_0 u + \delta_0\sqrt{2\epsilon}\,\gamma \tag{4.3.4}$$

where $\gamma(t) = \frac{dW}{dt}$ denotes the standard zero-mean Gaussian white noise characterized by

$$\langle\gamma(t)\rangle = 0,\ \langle\gamma(t_1)\gamma(t_2)\rangle = \delta(t_1 - t_2), \tag{4.3.5}$$

where $\langle\cdot\rangle$ represents the average over the ensemble of the stochastic process.

Equation (4.3.1) admits a single non-zero steady-state solution:

$$\overline{N} = \frac{a}{b}. \tag{4.3.6}$$

For the behaviour of the system (4.3.2) near the steady-state solution (4.3.6), we put

$$x = N - \overline{N}. \tag{4.3.7}$$

Then, Eq. (4.3.2) reduces to

$$\frac{dx}{dt} = -ax - bx^2 + u(t). \tag{4.3.8}$$

From (4.3.4) and (4.3.8), we obtain the following Itô stochastic differential equations:

$$\begin{aligned} dx_1 &= x_2 dt, \\ dx_2 &= m(x_1, x_2)dt + \sigma dW(t), \end{aligned} \tag{4.3.9}$$

where $x_1 = x$, $x_2 = \frac{dx}{dt}$ and $W(t)$ is a unit Wiener process.

The coefficient functions m and σ are given by

$$\begin{aligned} m(x_1, x_2) &= -(a + 2bx_1 + \delta_0)x_2 - a\delta_0 x_1 - b\delta_0 x_1^2 \\ \sigma &= \delta_0\sqrt{2\epsilon}. \end{aligned} \tag{4.3.10}$$

4.3.2 *Complex Stochastic Averaging*

We first convert the Markov vector (x_1, x_2) to another Markov vector (A, Φ), where the amplitudes $A(t)$ and the phases $\Phi(t)$ are slowly varying functions of time (for instants which are sufficiently remote from the initial time) and

$$\begin{aligned} x_1(t) &= A(t)\cos(\omega t + \Phi(t)), \\ x_2(t) &= -A(t)\omega\sin(\omega t + \Phi(t))). \end{aligned} \tag{4.3.11}$$

Solving for $A(t)$ and $\Phi(t)$:

$$A^2(t) = x_1^2 + \frac{x_2^2}{\omega^2}, \quad \Phi(t) = \tan^{-1}\left(\frac{x_2}{\omega x_1}\right) - \omega t. \tag{4.3.12}$$

Ariaratnam and Tam (1977) suggested an equivalent transformation:

$$x_1(t) = Z(t)e^{i\omega t} + \overline{Z(t)}e^{-i\omega t},$$
$$x_2(t) = i\omega\left\{Z(t)e^{i\omega t} - \overline{Z(t)}e^{-i\omega t}\right\}, \quad (i = \sqrt{-1}), \tag{4.3.13}$$

in which $Z(t)$ is complex random process and $\overline{Z(t)}$ is corresponding complex conjugate.

The two transformations (4.3.11) and (4.3.13) are related by

$$|Z|^2 = \frac{1}{4}A^2, \ \mathrm{Re}(Z) = \frac{1}{2}A\cos\Phi, \ \mathrm{Im}(Z) = \frac{1}{2}A\sin\Phi. \tag{4.3.14}$$

From (4.3.9) and (4.3.13), we obtain the following time averaged Itô equations (Bruckner and Lin 1987):

$$\mathrm{d}Z = m_z\mathrm{d}t + \sigma_{z1}\mathrm{d}W_1 + \sigma_{z2}\mathrm{d}W_2$$
$$\mathrm{d}\overline{Z} = \overline{m_z}\mathrm{d}t + \overline{\sigma_{z1}}\mathrm{d}W_1 + \overline{\sigma_{z2}}\mathrm{d}W_2 \tag{4.3.15}$$

($W_1(t)$ and $W_2(t)$ are two independent unit Wiener processes) with

$$m_z = -\frac{i}{2}\left\{\omega Z + \frac{1}{\omega}\langle m[x_1(Z,\overline{Z}), x_2(Z,\overline{Z})]e^{-i\omega t}\rangle_t\right\},$$
$$|\sigma_{z1}|^2 + |\sigma_{z2}|^2 = \frac{1}{4\omega^2}\langle\sigma^2\rangle_t. \tag{4.3.16}$$

$$\therefore m_z = -\frac{1}{2}\left\{i\omega Z + (a+\delta_0)Z - \frac{a\delta_0}{\omega}iZ\right\}, \ |\sigma_{z1}|^2 + |\sigma_{z2}|^2 = \frac{\epsilon\delta_0^2}{2\omega^2}. \tag{4.3.17}$$

It can be shown that the differential equation for the second-order moment $\langle|Z|^2\rangle$ is given by Bruckner and Lin (1987):

$$\frac{\mathrm{d}}{\mathrm{d}t}\langle|Z|^2\rangle = \langle Z\overline{m_z} + \overline{Z}m_z\rangle + \langle|\sigma_{z1}|^2 + |\sigma_{z2}|^2\rangle. \tag{4.3.18}$$

$$\therefore \langle A^2\rangle = D_0e^{-(a+\delta_0)t} + \frac{2\epsilon\delta_0^2}{\omega^2(a+\delta_0)}\left\{1 - e^{-(a+\delta_0)t}\right\}, \text{ where } D_0 = \langle A^2\rangle_{t=0}. \tag{4.3.19}$$

Therefore,

$$\lim_{t\to\infty}\langle A^2\rangle = \frac{2\epsilon\delta_0^2}{\omega^2(a+\delta_0)}. \tag{4.3.20}$$

From (4.3.19), we have

$$\frac{\mathrm{d}}{\mathrm{d}t}\langle A^2\rangle = \left\{\frac{2\epsilon\delta_0^2}{\omega^2} - (a+\delta_0)D_0\right\}\mathrm{e}^{-(a+\delta_0)t}. \tag{4.3.21}$$

1. If $\frac{2\epsilon\delta_0^2}{\omega^2} - (a+\delta_0)D_0 > 0$, i.e. $\langle A^2\rangle_{t=0} < \lim_{t\to\infty}\langle A^2\rangle$, then $\langle A^2\rangle$ is an increasing function of t.
2. If $\frac{2\epsilon\delta_0^2}{\omega^2} - (a+\delta_0)D_0 < 0$, i.e. $\langle A^2\rangle_{t=0} > \lim_{t\to\infty}\langle A^2\rangle$, then $\langle A^2\rangle$ is a decreasing function of t.
3. If $\frac{2\epsilon\delta_0^2}{\omega^2} - (a+\delta_0)D_0 = 0$, i.e. $\langle A^2\rangle_{t=0} = \lim_{t\to\infty}\langle A^2\rangle$, then $\langle A^2\rangle$ is independent of time t.

Now, in the deterministic environment, the eigen value of the 1×1 interaction matrix is $-a$, which is negative, showing that the equilibrium point is stable. In the stochastic environment, whose random fluctuation has intensity ϵ, the stability provided by the population interaction dynamics is again characterized by a. It is no longer enough that $a > 0$, for if $\epsilon \gg a$, population exhibits large fluctuation, which rapidly lead to extinction, because in this case there is no internal mechanism to reestablish the stability once it is perturbed by internal noise. For $a > \epsilon$, but not much greater, population are likely to undergo significant fluctuations, even though they persist for long times. Finally, if $a \gg \epsilon$, population fluctuations are relatively small, and the environment is effectively deterministic (in this case deterministic solution is recovered). It is also noted that as $\delta_0 \to \infty$, the colored noise $u(t)$ becomes the white noise and the system will be unstable which is the well-known result.

4.3.3 Change in Net Growth Rate

In this section, we have studied the effects of randomly fluctuating environment on logistic growth model. In the absence of fluctuating environment, the population is assumed to grow according to Eq. (4.3.1), i.e. according to

$$\frac{\mathrm{d}N}{\mathrm{d}t} = aN - bN^2.$$

The fluctuating environment may affect the growth in several different ways, one of which is changes in net growth rate parameter. The fluctuating environment may introduce a stochastic element in the growth rate a. A stochastic form of a may be

$$a = a_0 + u(t), \; a_0 > 0, \tag{4.3.22}$$

where the random perturbation term $u(t)$ is assumed to be a *colored noise* or *Ornstein–Uhlenbeck process* which is more realistic noise than *white noise*. The mathematical expectation and correlation function of the process $u(t)$ are given by (4.3.3). The stochastic differential equation describing the growth of the population is then

$$\frac{dN}{dt} = (a_0 + u)N - bN^2 = a_0N - bN^2 + uN. \tag{4.3.23}$$

For the behaviour of the system (4.3.23) near the steady-state solution (4.3.6), we put $x = N - \overline{N}$ and so Eq. (4.3.23) reduces to the form:

$$\frac{dx}{dt} = -a_0x - bx^2 + (x + N)u(t). \tag{4.3.24}$$

From (4.3.4) and (4.3.24), we obtain the following Itô stochastic differential equations:

$$\begin{aligned} dx_1 &= x_2 dt, \\ dx_2 &= m(x_1, x_2)dt + \sigma dW(t), \end{aligned} \tag{4.3.25}$$

where $x_1 = x$, $x_2 = \dfrac{dx}{dt}$ and $W(t)$ is a unit Wiener process.

The coefficient functions m and σ are given by

$$\begin{aligned} m(x_1, x_2) &= -(a + 2bx_1 + \delta_0)x_2 + \frac{x_2^2}{x_1 + \overline{N}} - a\delta_0x_1 - b\delta_0x_1^2 \\ &\quad + \frac{(ax_1 + bx_1^2)x_2}{x_1 + \overline{N}}, \text{ and } \sigma = (x_1 + \overline{N})\delta_0\sqrt{2\epsilon}. \end{aligned} \tag{4.3.26}$$

We convert the Markov vector (x_1, x_2) to another Markov vector (A, Φ), where the amplitudes $A(t)$ and the phases $\Phi(t)$ are slowly varying functions of time (for instants which are sufficiently remote from the initial time) and

$$x_1(t) = A(t)\cos(\omega t + \Phi(t)),$$

$$x_2(t) = -A(t)\omega\sin(\omega t + \Phi(t))).$$

Proceeding as in Sect. (4.3.2), we obtain

$$\frac{d}{dt}\langle A^2\rangle + \left\{(a + \delta_0) - \frac{\epsilon\delta_0^2}{\omega^2}\right\}\langle A^2\rangle = \frac{\overline{N}^2\epsilon\delta_0^2}{2\omega^2}. \tag{4.3.27}$$

$$\begin{aligned} \therefore \langle A^2\rangle &= D_0\exp\left\{-\left((a + \delta_0) - \frac{\epsilon\delta_0^2}{\omega^2}\right)t\right\} + \frac{a^2\epsilon\delta_0^2}{2b^2\{(a + \delta_0)\omega^2 - \epsilon\delta_0^2\}} \\ &\quad \times\left[1 - \exp\left\{-\left((a + \delta_0) - \frac{\epsilon\delta_0^2}{\omega^2}\right)t\right\}\right], \text{ where } D_0 = \langle A^2\rangle_{t=0}. \end{aligned} \tag{4.3.28}$$

If we choose $\omega^2 = \delta_0^2$, then

$$\langle A^2 \rangle = D_0 e^{-(a+\delta_0-\epsilon)t} + \frac{a^2\epsilon}{2b^2(a+\delta_0-\epsilon)}\{1 - e^{-(a+\delta_0-\epsilon)t}\}. \tag{4.3.29}$$

From (4.3.29), we have

$$\frac{d}{dt}\langle A^2 \rangle = \left\{ \frac{a^2\epsilon}{2b^2} - (a+\delta_0-\epsilon)D_0 \right\} e^{-(a+\delta_0-\epsilon)t}. \tag{4.3.30}$$

1. If $\frac{a^2\epsilon}{2b^2} - (a+\delta_0-\epsilon)D_0 > 0$, i.e. $D_0 < \frac{a^2\epsilon}{2b^2(a+\delta_0-\epsilon)}$, then $\langle A^2 \rangle$ is an increasing function of t.
2. If $\frac{a^2\epsilon}{2b^2} - (a+\delta_0-\epsilon)D_0 < 0$, i.e. $D_0 > \frac{a^2\epsilon}{2b^2(a+\delta_0-\epsilon)}$, then $\langle A^2 \rangle$ is a decreasing function of t.
3. If $\frac{a^2\epsilon}{2b^2} - (a+\delta_0-\epsilon)D_0 = 0$, i.e. $D_0 = \frac{a^2\epsilon}{2b^2(a+\delta_0-\epsilon)}$, then $\langle A^2 \rangle$ is independent of time t.

In the deterministic environment, the steady-state solution $\overline{N}$ is stable since $a > 0$. In the stochastic environment, whose random fluctuation has intensity ϵ, the stability provided by the population interaction dynamics is again characterized by a. It is no longer enough that $a > 0$, for if $\epsilon \gg a + \delta_0$, where δ_0^{-1} is the correlation time of the colored noise $u(t)$, population exhibits large fluctuation, which rapidly lead to extinction, because in this case there is no internal mechanism to reestablish the stability once it is perturbed by the environmental noise. For $a + \delta_0 > \epsilon$, but not much greater, population are likely to undergo significant fluctuations, even though they persist for long times. Finally, if $a \gg \epsilon$, population fluctuations are relatively small, and the environment is effectively deterministic (in this case deterministic solution is recovered).

References

Ariaratnam, S.T., Tam, D.S.F.: Moment stability of coupled linear systems under combined harmonics and stochastic excitation. In: Clarkson, B.L. (ed.) Stochastic Problems in Dynamics. Pitman, London (1977)

Bruckner, A., Lin, Y.K.: Application of complex stochastic averaging to non-linear random vibration problems. Int. J. Non-Linear Mech. **22**, 237–250 (1987)

Cai, G.Q., Lin, Y.K.: On exact stationary solutions of equivalent non-linear stochastic systems. Int. J. Non-Linear Mech. **23**(4), 315–325 (1988)

Uhlenbeck, G.E., Ornstein, L.S.: On the theory of Brownian motion. In: Wax, N. (ed.) Selected Papers on Noise and Stochastic Process. Dover, New York (1954)

Chapter 5
Stability Behaviour in Randomly Fluctuating Versus Deterministic Environments of Two Interacting Species

5.1 Introduction

It is a known fact that most natural phenomena do not follow strictly deterministic laws but rather oscillate randomly about some average behaviour, which is especially true in the context of ecological models where environmental influences should be taken as stochastic (fluctuating). Volterra–Lotka model of interacting species is taken as a starting point in many models of theoretical ecology. The influence of the environment has to be taken into account by arbitrarily augmenting the deterministic equations with stochastic terms or taken the environmental parameters as time-dependent and rapidly varying. In both situations, the corresponding stochastic properties have to be postulated on the basis of experimental evidence.

5.2 Stability Behaviour in Randomly Fluctuating Versus Deterministic Environments of the Gomatam Model of Interacting Species

Gomatam (1974) has suggested a deterministic model for interacting species which, by substituting Gompertzian interactions for Verhulstian ones in the Volterra equations, become exactly solvable while retaining the broad features of the Volterra model. In this article, we have studied the relationship between stability behaviours of this model in which environmental parameters are prescribed constants (deterministic environment) and the corresponding model in which these parameters have an element of random fluctuation (stochastic environment).

© Springer Nature Singapore Pte Ltd. 2021

G. Samanta, *Deterministic, Stochastic and Thermodynamic Modelling of some Interacting Species*, Forum for Interdisciplinary Mathematics,
https://doi.org/10.1007/978-981-16-6312-3_5

5.2.1 Gomatam Model: Basic Stochastic Differential Equations

For Gomatam model, the prey population N_1 and the predator population N_2 satisfy the following deterministic equations:

$$\begin{aligned}\frac{dN_1}{dt} &= \alpha_{10}N_1 - \beta_1 N_1 \log N_2 - \gamma_1 N_1 \log N_1 \\ \frac{dN_2}{dt} &= -\alpha_{20}N_2 + \beta_2 N_2 \log N_1 - \gamma_2 N_2 \log N_2\end{aligned} \tag{5.2.1}$$

where $\alpha_{10}, \alpha_{20} > 0$ and $\beta_1, \gamma_1, \beta_2, \gamma_2 > 0$ denote the growth rates and interaction coefficients, respectively, which are assumed to be constants.

Here, we are considering following two cases.

Case I: Deterministic Environment

In a deterministic environment, we have constant environmental parameters α_{10}, α_{20}. Letting $x_i = \log N_i$, Eq. (5.2.1) can be rewritten as

$$\begin{aligned}\frac{dx_1}{dt} &= \alpha_{10} - \gamma_1 x_1 - \beta_1 x_2 \\ \frac{dx_2}{dt} &= -\alpha_{20} + \beta_2 x_1 - \gamma_2 x_2\end{aligned} \tag{5.2.2}$$

This system has a unique non-trivial equilibrium at the point

$$x_1^* = \frac{\alpha_{10}\gamma_2 + \alpha_{20}\beta_1}{\gamma_1\gamma_2 + \beta_1\beta_2}, \quad x_2^* = \frac{\alpha_{10}\beta_2 - \alpha_{20}\gamma_1}{\gamma_1\gamma_2 + \beta_1\beta_2} \tag{5.2.3}$$

It is immediately apparent that in the absence of predators, the limit value of the prey population will be $x_1' = \frac{\alpha_{10}}{\gamma_1}$ and in the absence of intraspecific competition among the predators, the limit value of the prey population will be $x_1'' = \frac{\alpha_{20}}{\beta_2}$. The realization of the obvious condition:

$$x_1' > x_1'' \text{ implies } \alpha_{10}\beta_2 - \alpha_{20}\gamma_1 > 0 \tag{5.2.4}$$

makes x_2^* positive, and hence, this equilibrium exists.

The 2×2 coefficient matrix of system (5.2.2) is

$$A = \begin{bmatrix} -\gamma_1 & -\beta_1 \\ \beta_2 & -\gamma_2 \end{bmatrix} \tag{5.2.5}$$

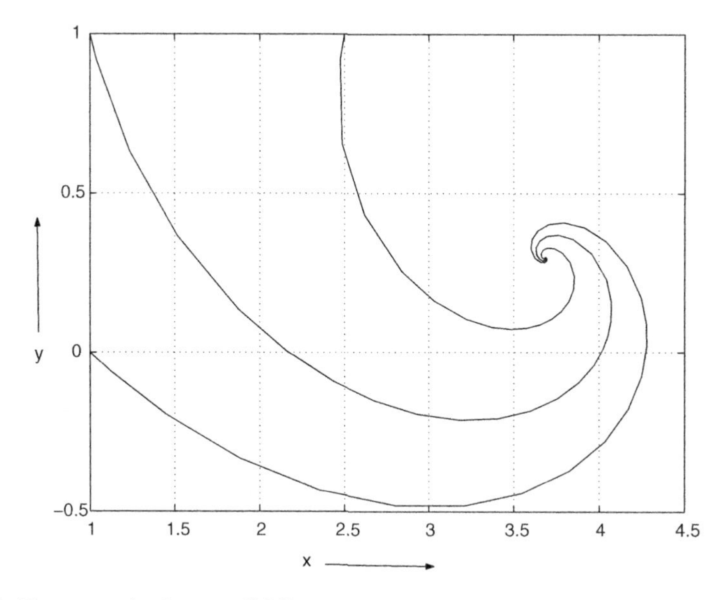

Fig. 5.1 Phase portrait of system (5.2.2)

The eigenvalues of this matrix are

$$\lambda_1 = \frac{-(\gamma_1 + \gamma_2) + \sqrt{\Delta}}{2}, \quad \lambda_2 = \frac{-(\gamma_1 + \gamma_2) - \sqrt{\Delta}}{2} \tag{5.2.6}$$

where

$$\Delta = (\gamma_1 + \gamma_2)^2 - 4(\gamma_1\gamma_2 + \beta_1\beta_2) \tag{5.2.7}$$

Both eigenvalues of this matrix have negative real parts. Therefore, the unique non-trivial equilibrium point of Gomatam model is asymptotically stable in a deterministic environment (Figs. 5.1 and 5.2).

Case II: Stochastic Environment

In a stochastic environment suppose that the environmental parameters α_{10}, α_{20} vary randomly, and write

$$\begin{aligned} \alpha_1 &= \alpha_{10} + D\,\gamma_1'(t) \\ -\alpha_2 &= -\alpha_{20} + D\,\gamma_2'(t) \end{aligned} \tag{5.2.8}$$

Here α_{10}, α_{20} are constants, being the mean values of α_1, α_2 respectively, and $\gamma_1'(t),\ \gamma_2'(t)$ are standard white noises characterized by

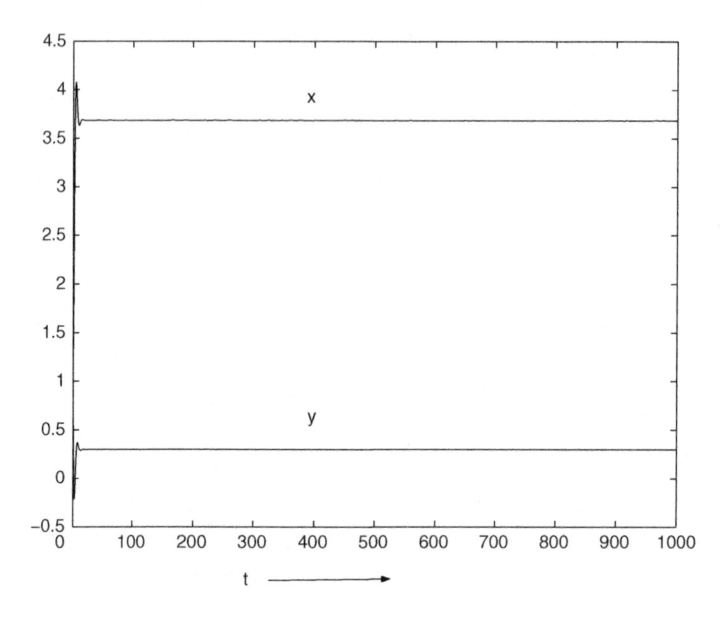

Fig. 5.2 Trajectories of system (5.2.2)

$$\begin{aligned}&\langle \gamma_i'(t)\rangle = 0, \quad \langle \gamma_i'(t_1)\gamma_i'(t_2)\rangle = \delta(t_1 - t_2)\\ \text{and } &\langle \gamma_i'(t_1)\gamma_j'(t_2)\rangle = \rho\, \delta(t_1 - t_2), (i \neq j)\ i, j = 1, 2\end{aligned} \tag{5.2.9}$$

Therefore, the system of equations (5.2.1) becomes

$$\begin{aligned}\frac{\mathrm{d}N_1}{\mathrm{d}t} &= \alpha_{10}N_1 - \beta_1 N_1 \log N_2 - \gamma_1 N_1 \log N_1 + DN_1\gamma_1'(t)\\ \frac{\mathrm{d}N_2}{\mathrm{d}t} &= -\alpha_{20}N_2 + \beta_2 N_2 \log N_1 - \gamma_2 N_2 \log N_2 + DN_2\gamma_2'(t)\end{aligned} \tag{5.2.10}$$

Though Gaussian white noises are so very irregular, these are extremely useful to model rapidly fluctuating phenomena. Of course true white noises do not occur in nature. However, as can be seen by studying their spectra, thermal noises in electrical resistance, the force acting on a Brownian particle, climate fluctuations, disregarding the periodicities of astronomical origin, etc. are white to a very good approximation. These examples support the usefulness of the white noise idealization in applications to natural systems. Furthermore, it can be proved that the process $(N_1, N_2)^{\mathrm{T}}$, solution of (5.2.10), is Markovian if and only if the external noises are white. These results explain the importance and appeal of the white noise idealization (Horsthemke and Lefever 1984).

It is noted that $\gamma_i'(t)$ are not defined in the ordinary sense. It can be proved that $\gamma_i'(t)$ are the derivatives of the Wiener process $W_i(t)$ in the generalized functions sense. Therefore to avoid the theory of generalized stochastic processes, we rewrite (5.2.10) in the form:

$$\begin{aligned} dN_1 &= (\alpha_{10}N_1 - \beta_1 N_1 \log N_2 - \gamma_1 N_1 \log N_1)dt + DN_1 dW_1(t) \\ dN_2 &= (-\alpha_{20}N_2 + \beta_2 N_2 \log N_1 - \gamma_2 N_2 \log N_2)dt + DN_2 dW_2(t) \end{aligned} \tag{5.2.11}$$

Here, $W_1(t)$ and $W_2(t)$ are two standard Wiener processes. Indeed in (5.2.11) ordinary processes appear. However, it can be shown that Eq. (5.2.11) as it stand have no meaning (Van Kampen 1981). In fact, there are only two reasonable interpretations of the stochastic differential equations (5.2.11), one is due to *K*. Itô (1942) and the other is due to Stratonovich (1963). Stratonovich interpretation conserves the ordinary rules of calculus, and in this case, the stochastic differential equations can be considered as an ensemble of ordinary differential equations. This suggests that if the stochastic differential equations with white noises are the limit of stochastic differential equations with more realistic noise processes, then the Stratonovich interpretation will be reasonable. On the other hand Itô, interpretation does not conserve the ordinary rules of calculus; however, this is not a serious drawback. The classical rules of calculus have no special intrinsic values; it just so happens that they are the only ones we are familiar with.

The Itô interpretation makes maximal use of the important property of the Wiener process of having independent increments. Furthermore, if the external noises are truly white, then this is the only reasonable choice from physical point of view (Horsthemke and Lefever 1984).

Stratonovich Interpretation

Since Stratonoivich interpretation conserves the ordinary rules of calculus, the system of equations (5.2.10) can be written as

$$\begin{aligned} \frac{dX_1}{dt} &= \alpha_{10} - \gamma_1 X_1 - \beta_1 X_2 + D\gamma_1'(t) \\ \frac{dX_2}{dt} &= -\alpha_{20} + \beta_2 X_1 - \gamma_2 X_2 + D\gamma_2'(t) \end{aligned} \tag{5.2.12}$$

where $X_1 = \log N_1$ and $X_2 = \log N_2$.
The system of equations (5.2.12) can be written in the matrix form:

$$dX = A'(X)dt + BdW(t) \tag{5.2.13}$$

where

$$X = \begin{bmatrix} X_1 \\ X_2 \end{bmatrix}, A' = \begin{bmatrix} \alpha_{10} - \gamma_1 X_1 - \beta_1 X_2 \\ -\alpha_{20} + \beta_2 X_1 - \gamma_2 X_2 \end{bmatrix}, B = \begin{bmatrix} D & 0 \\ 0 & D \end{bmatrix} \text{ and } \mathrm{d}W = \begin{bmatrix} \mathrm{d}W_1 \\ \mathrm{d}W_2 \end{bmatrix} \tag{5.2.14}$$

Itô Interpretation

Equation (5.2.11) can be written in the matrix form:

$$\mathrm{d}N = b(N)\mathrm{d}t + \sigma \mathrm{d}W(t) \tag{5.2.15}$$

where

$$\begin{aligned} N &= \begin{bmatrix} N_1 \\ N_2 \end{bmatrix}, b = \begin{bmatrix} \alpha_{10} N_1 - \gamma_1 N_1 \log N_1 - \beta_1 N_1 \log N_2 \\ -\alpha_{20} N_2 + \beta_2 N_2 \log N_1 - \gamma_2 N_2 \log N_2 \end{bmatrix}, \\ \sigma &= \begin{bmatrix} DN_1 & 0 \\ 0 & DN_2 \end{bmatrix} \text{ and } \mathrm{d}W = \begin{bmatrix} \mathrm{d}W_1 \\ \mathrm{d}W_2 \end{bmatrix} \end{aligned} \tag{5.2.16}$$

Let $X_1 = \log N_1$ and $X_2 = \log N_2$.
Then using Itô formula, we have (Schuss 1980)

$$\mathrm{d}X_1 = \left(\frac{\partial X_1}{\partial t} + b^{\mathrm{T}} \cdot \nabla_N X_1 + \frac{1}{2} \sum_{i,j} a_{ij} \frac{\partial^2 X_1}{\partial N_i \partial N_j} \right) \mathrm{d}t + \nabla_N X_1^{\mathrm{T}} \sigma \mathrm{d}W \tag{5.2.17}$$

where $a_{ij} = (\sigma \sigma^{\mathrm{T}})_{ij}$.

$$\text{Therefore, } \mathrm{d}X_1 = \left\{ \frac{1}{N_1} (\alpha_{10} N_1 - \gamma_1 N_1 \log N_1 - \beta_1 N_1 \log N_2) - \frac{D^2}{2} \right\} \mathrm{d}t + D\mathrm{d}W_1 \tag{5.2.18}$$

Similarly,

$$\mathrm{d}X_2 = \left\{ \frac{1}{N_2} (-\alpha_{20} N_2 - \gamma_2 N_2 \log N_2 + \beta_2 N_2 \log N_1) - \frac{D^2}{2} \right\} \mathrm{d}t + D\mathrm{d}W_2 \tag{5.2.19}$$

$$\begin{aligned} \text{Therefore, } \frac{\mathrm{d}X_1}{\mathrm{d}t} &= \left(\alpha_{10} - \frac{D^2}{2} \right) - \gamma_1 X_1 - \beta_1 X_2 + D\gamma_1'(t) \\ \frac{\mathrm{d}X_2}{\mathrm{d}t} &= \left(-\alpha_{20} - \frac{D^2}{2} \right) + \beta_2 X_1 - \gamma_2 X_2 + D\gamma_2'(t) \end{aligned} \tag{5.2.20}$$

The system of equations (5.2.20) can be written in the matrix form:

$$\mathrm{d}X = A''(X)\mathrm{d}t + B\mathrm{d}W(t) \tag{5.2.21}$$

where

$$X = \begin{bmatrix} X_1 \\ X_2 \end{bmatrix}, A'' = \begin{bmatrix} \left(\alpha_{10} - \frac{D^2}{2}\right) - \gamma_1 X_1 - \beta_1 X_2 \\ \left(-\alpha_{20} - \frac{D^2}{2}\right) + \beta_2 X_1 - \gamma_2 X_2 \end{bmatrix}, B = \begin{bmatrix} D & 0 \\ 0 & D \end{bmatrix} \text{ and } dW(t) = \begin{bmatrix} dW_1(t) \\ dW_2(t) \end{bmatrix} \quad (5.2.22)$$

The Fokker–Planck equation for the probability density $p(x_1, x_2, t)$ is (Gardiner 1983)

$$\partial_t p = -\sum_i \partial_i [A_i(x) p] + \frac{1}{2} \sum_{i,j} \partial_i \partial_j \{[BB^{\mathrm{T}}]_{ij} p\} \quad (5.2.23)$$

Therefore, the stationary probability density $p_s(x_1, x_2)$ is given by

$$-\sum_i \partial_i [A_i(x) p_s] + \frac{1}{2} \sum_{i,j} \partial_i \partial_j \{[BB^{\mathrm{T}}]_{ij} p_s\} = 0 \quad (5.2.24)$$

After some simplifications (Risken 1984), we have obtained:

$$p_s(x_1, x_2) = \frac{1}{2\pi\sigma_1\sigma_2\sqrt{1-\rho_1^2}} \exp\left[-\frac{1}{2(1-\rho_1^2)} \left\{ \frac{(x_1-\mu_1)^2}{\sigma_1^2} - 2\rho_1 \frac{(x_1-\mu_1)(x_2-\mu_2)}{\sigma_1\sigma_2} + \frac{(x_2-\mu_2)^2}{\sigma_2^2} \right\}\right], \quad -\infty < x_1, x_2 < \infty \quad (5.2.25)$$

$$\text{where } \sigma_1 = D\left\{\frac{(\beta_1-\gamma_2)^2 + 2\beta_1\gamma_2(1-\rho) + \gamma_1\gamma_2 + \beta_1\beta_2}{2(\gamma_1+\gamma_2)(\gamma_1\gamma_2+\beta_1\beta_2)}\right\}^{\frac{1}{2}}, \quad (5.2.26)$$

$$\sigma_2 = D\left\{\frac{(\beta_2-\gamma_1)^2 + 2\beta_2\gamma_1(1+\rho) + \gamma_1\gamma_2 + \beta_1\beta_2}{2(\gamma_1+\gamma_2)(\gamma_1\gamma_2+\beta_1\beta_2)}\right\}^{\frac{1}{2}} \quad (5.2.27)$$

$$\text{and } \rho_1^2 = \frac{(\beta_2\gamma_2 - \beta_1\gamma_1 + 2\rho\gamma_1\gamma_2)^2}{\{(\beta_1-\gamma_2)^2 + 2\beta_1\gamma_2(1-\rho) + \gamma_1\gamma_2 + \beta_1\beta_2\}\{(\beta_2-\gamma_1)^2 + 2\beta_2\gamma_1(1+\rho) + \gamma_1\gamma_2 + \beta_1\beta_2\}} \quad (5.2.28)$$

Therefore, we observe that if $0 \le \rho_1^2 < 1$, then the stationary probability density $p_s(x_1, x_2)$ exists. If $\gamma_1'(t),\ \gamma_2'(t)$ are independent standard white noises, then

$$0 \le \rho_1^2 = \frac{(\beta_2\gamma_2 - \beta_1\gamma_1)^2}{\{\beta_1^2 + \gamma_2^2 + \gamma_1\gamma_2 + \beta_1\beta_2\}\{\beta_2^2 + \gamma_1^2 + \gamma_1\gamma_2 + \beta_1\beta_2\}} < 1 \quad (5.2.28a)$$

Therefore, the stationary probability density $p_s(x_1, x_2)$ exists if $\gamma_1'(t),\ \gamma_2'(t)$ are independent standard white noises.

For system (5.2.12),

$$\mu_1 = \frac{\alpha_{10}\gamma_2 + \alpha_{20}\beta_1}{\gamma_1\gamma_2 + \beta_1\beta_2}, \quad \mu_2 = \frac{\alpha_{10}\beta_2 - \alpha_{20}\gamma_1}{\gamma_1\gamma_2 + \beta_1\beta_2} \tag{5.2.29}$$

For system (5.2.20),

$$\mu_1 = \frac{(\alpha_{10} - \frac{D^2}{2})\gamma_2 - (-\alpha_{20} - \frac{D^2}{2})\beta_1}{\gamma_1\gamma_2 + \beta_1\beta_2}, \quad \mu_2 = \frac{(\alpha_{10} - \frac{D^2}{2})\beta_2 + (-\alpha_{20} - \frac{D^2}{2})\gamma_1}{\gamma_1\gamma_2 + \beta_1\beta_2} \tag{5.2.30}$$

By integrating over all values of x_2, we obtain

$$p_s(x_1) = \frac{1}{\sigma_1\sqrt{2\pi}}\exp\left\{\frac{-(x_1 - \mu_1)^2}{2\sigma_1^2}\right\} \quad (-\infty < x_1 < \infty) \tag{5.2.31}$$

$$\text{Therefore,} \quad p_s(n_1) = \frac{1}{n_1\sigma_1\sqrt{2\pi}}\exp\left\{\frac{-(\log n_1 - \mu_1)^2}{2\sigma_1^2}\right\}, \quad n_1 > 0$$
$$p_s(n_1) = 0, \quad n_1 = 0 \tag{5.2.32}$$

It can be shown that

$$< N_1^r >= \exp\left\{r\mu_1 + \frac{1}{2}r^2\sigma_1^2\right\} \tag{5.2.33}$$

Therefore,

$$< (N_1 - < N_1 >)^2 >= e^{(2\mu_1+\sigma_1^2)}\{e^{\sigma_1^2} - 1\} \tag{5.2.34}$$

Similarly,

$$< (N_2 - < N_2 >)^2 >= e^{(2\mu_2+\sigma_2^2)}\{e^{\sigma_2^2} - 1\} \tag{5.2.35}$$

where σ_1, σ_2 are given by (5.2.26) and (5.2.27), respectively. For Stratonovich interpretation, μ_1, μ_2 are given by (5.2.29), and for Itô interpretation, μ_1, μ_2 are given by (5.2.30).

5.2.2 Discussion

In this article, we have studied the relationship between stability behaviours of the Gomatam model for interacting species (by substituting Gompertzian interactions for Verhulstian ones in the Volterra equations) in which environmental parameters are prescribed constants (deterministic environment) and the corresponding model in which these parameters have an element of random fluctuation (stochastic environment). In the stochastic environment, the environmental parameters α_{10}, α_{20} have been perturbed by Gaussian white noises. In the deterministic environment: the eigenvalues of the interaction matrix of the underlying system are $\lambda_{1,2} = \frac{-(\gamma_1+\gamma_2)\pm\sqrt{\Delta}}{2}$, where $\Delta = (\gamma_1 + \gamma_2)^2 - 4(\gamma_1\gamma_2 + \beta_1\beta_2)$. Therefore, $(\gamma_1 + \gamma_2)$ can be treated as the stability determining quantity and the deterministic stability criteria is fulfilled (since $(\gamma_1 + \gamma_2) > 0$). In the stochastic (fluctuating) environmental conditions: $(\gamma_1 + \gamma_2)$ can also be treated as the stability determining quantity if $0 \leq \rho_1^2 < 1$ (where ρ_1^2 is given by Eq. 5.2.28). But in this situation, it is no longer enough that $(\gamma_1 + \gamma_2) > 0$, because if $D^2 \gg (\gamma_1 + \gamma_2)$, population exhibits large fluctuations leading to rapid extinction (by Eqs. 5.2.26, 5.2.27, 5.2.34 and 5.2.35). It is remembered that D^2 is the intensity of the environmental noise. When D^2 and $(\gamma_1 + \gamma_2)$ are commensurate: the system is likely to undergo significant fluctuations, even though they persist for long times. For the situations $D^2 \ll (\gamma_1 + \gamma_2)$: population fluctuations are relatively small, and so, the environment is effectively deterministic (non-fluctuating). These results agree with May (1973).

5.3 Stability Behaviour of the Gomatam Model of Interacting Species in a Randomly Fluctuating Environment

In this article, we have studied the relationship between stability behaviours of the Gomatam model in which environmental parameters are prescribed constants (deterministic environment) and the corresponding model in which these parameters have an element of random fluctuation perturbed by wideband stationary stochastic process with zero mean (stochastic environment).

5.3.1 Gomatam Model: Modified Stochastic Differential Equations

For Gomatam model, the prey population N_1 and the predator population N_2 satisfy the deterministic differential equations:

$$
\begin{aligned}
\frac{dN_1}{dt} &= \alpha_{10}N_1 - \beta_1 N_1 \log N_2 - \gamma_1 N_1 \log N_1 \\
\frac{dN_2}{dt} &= -\alpha_{20}N_2 + \beta_2 N_2 \log N_1 - \gamma_2 N_2 \log N_2
\end{aligned}
\qquad (5.3.1)
$$

where $\alpha_{10}, \alpha_{20} > 0$ and $\beta_1, \gamma_1, \beta_2, \gamma_2 > 0$ denote the growth rates and interaction coefficients, respectively, which are assumed to be constants.

Letting $X_i = \log N_i$, Eq. (5.3.1) can be rewritten as

$$\begin{aligned}\frac{\mathrm{d}x_1}{\mathrm{d}t} &= \alpha_{10} - \gamma_1 x_1 - \beta_1 x_2 \\ \frac{\mathrm{d}x_2}{\mathrm{d}t} &= -\alpha_{20} + \beta_2 x_1 - \gamma_2 x_2\end{aligned} \tag{5.3.2}$$

This system has a unique non-trivial equilibrium at the point

$$x_1^* = \frac{\alpha_{10}\gamma_2 + \alpha_{20}\beta_1}{\gamma_1\gamma_2 + \beta_1\beta_2}, \quad x_2^* = \frac{\alpha_{10}\beta_2 - \alpha_{20}\gamma_1}{\gamma_1\gamma_2 + \beta_1\beta_2} \tag{5.3.3}$$

It is immediately apparent that in the absence of predators, the limit value of the prey population will be $x_1' = \frac{\alpha_{10}}{\gamma_1}$, and in the absence of intraspecific competition among the predators, the limit value of the prey population will be $x_1'' = \frac{\alpha_{20}}{\beta_2}$. The realization of the obvious condition:

$$x_1' > x_1'' \text{ implies } \alpha_{10}\beta_2 - \alpha_{20}\gamma_1 > 0 \tag{5.3.4}$$

makes x_2^* positive, and hence, this equilibrium exists.

Putting $X_1 = x_1 - x_1^*$ and $X_2 = x_2 - x_2^*$, the system (5.3.2) can be rewritten as:

$$\begin{aligned}\frac{\mathrm{d}X_1}{\mathrm{d}t} &= -\gamma_1 X_1(t) - \beta_1 X_2(t) \\ \frac{\mathrm{d}X_2}{\mathrm{d}t} &= \beta_2 X_1(t) - \gamma_2 X_2(t)\end{aligned} \tag{5.3.5}$$

The modified stochastic version of system (5.3.5) to be considered is:

$$\begin{aligned}\frac{\mathrm{d}X_1}{\mathrm{d}t} &= -\gamma_1(t) X_1(t) - \beta_1(t) X_2(t) \\ \frac{\mathrm{d}X_2}{\mathrm{d}t} &= \beta_2(t) X_1(t) - \gamma_2(t) X_2(t)\end{aligned} \tag{5.3.6}$$

where the stochastic parameters $\beta_i(t), \gamma_i(t), \ (i = 1, 2)$ have been expressed as:

$$\begin{aligned}\beta_i(t) &= \beta_{i0} + \varepsilon^2\beta_{i1} + \varepsilon f(t)\beta_{i2} \\ \gamma_i(t) &= \gamma_{i0} + \varepsilon^2\gamma_{i1} + \varepsilon f(t)\gamma_{i2}.\end{aligned} \tag{5.3.7}$$

where $\beta_{i0}, \beta_{i1}, \beta_{i2}, \gamma_{i0}, \gamma_{i1}, \gamma_{i2}$ are non-negative constants, $0 < \varepsilon << 1$ and $f(t)$ is a wideband stationary stochastic process with zero mean.

Then the behaviour of the stochastic system (5.3.6) is given in the following theorem.

Theorem 5.3.1 *If* $|\gamma_{10} - \gamma_{20}| < 2(\beta_{10}\beta_{20})^{1/2}$ *and the correlation time of the stochastic process* $f(t)$ *is sufficiently small, then the system (5.3.6) is stable or unstable (in mean square) according as* $\Omega \leq 0$ *or* $\Omega > 0$*, respectively, where*

$$\begin{aligned}
\Omega &= \frac{1}{\omega^2}[\{\gamma_{12}\gamma_{22}(\gamma_{10} - \gamma_{20})^2 + (\gamma_{12} + \gamma_{22})^2(3\beta_{10}\beta_{20} - (\gamma_{10} - \gamma_{20})^2) \\
&\quad +(\beta_{10}\beta_{22} + \beta_{12}\beta_{20})((\gamma_{22} - \gamma_{12})(\gamma_{10} + \gamma_{20}) - \beta_{10}\beta_{22} - \beta_{20}\beta_{12})\}S(0) \\
&\quad +3\{\beta_{10}\beta_{20}(\gamma_{22} - \gamma_{12})^2 + (\beta_{10}\beta_{22} - \beta_{20}\beta_{12})^2 \\
&\quad +(\gamma_{10} + \gamma_{20})(\gamma_{12} - \gamma_{22})(\beta_{10}\beta_{22} + \beta_{12}\beta_{20}) \\
&\quad +\beta_{12}\beta_{22}(\gamma_{10} - \gamma_{20})^2\}S(\omega) - (\gamma_{11} + \gamma_{21})], \\
\omega^2 &= 4\beta_{10}\beta_{20} - (\gamma_{10} - \gamma_{20})^2
\end{aligned}$$

and $S(\omega)$ *is the spectral density function of* $f(t)$*. Also* $\Omega < 0$ *guarantees asymptotic stability of the system with probability one.*

Proof The system of equation (5.3.6) can be rewritten in the matrix form:

$$\frac{dX}{dt} = [A^{(0)} + \varepsilon^2 B^{(0)} + \varepsilon f(t)P^{(0)}]X \tag{5.3.8}$$

where

$$X = \begin{bmatrix} X_1 \\ X_2 \end{bmatrix}, \quad A^{(0)} = \begin{bmatrix} -\gamma_{10} & -\beta_{10} \\ \beta_{20} & -\gamma_{20} \end{bmatrix},$$

$$B^{(0)} = \begin{bmatrix} -\gamma_{11} & -\beta_{11} \\ \beta_{21} & -\gamma_{21} \end{bmatrix} \text{ and } P^{(0)} = \begin{bmatrix} -\gamma_{12} & -\beta_{12} \\ \beta_{22} & -\gamma_{22} \end{bmatrix}.$$

Therefore, system (5.3.8) is a weakly stochastic system governed by linear state equations. These types of state equations also include the class of parametrically excited mechanical systems (Ariaratnam and Tam 1974).

The eigenvalues of the matrix $A^{(0)}$ are as follows:

$$\lambda_1 = \frac{-(\gamma_{10} + \gamma_{20}) + i\omega}{2} \text{ and } \lambda_2 = \frac{-(\gamma_{10} + \gamma_{20}) - i\omega}{2}$$

where

$$\omega^2 = 4\beta_{10}\beta_{20} - (\gamma_{10} - \gamma_{20})^2,$$

which is positive according to the assumption

$$|\gamma_{10} - \gamma_{20}| < 2(\beta_{10}\beta_{20})^{1/2}.$$

Let $\begin{bmatrix} x_1 \\ x_2 \end{bmatrix}$ and $\begin{bmatrix} x_1' \\ x_2' \end{bmatrix}$ be two eigenvectors of $A^{(0)}$ where

$$\begin{aligned}
x_1 &= \beta_{10}/\left\{\beta_{10}^2 + (\gamma_{10} + \lambda_1)^2\right\}^{1/2}, \\
x_2 &= -(\gamma_{10} + \lambda_1)/\left\{\beta_{10}^2 + (\gamma_{10} + \lambda_1)^2\right\}^{1/2}, \\
x_1' &= \beta_{10}/\left\{\beta_{10}^2 + (\gamma_{10} + \lambda_2)^2\right\}^{1/2}, \\
x_2' &= -(\gamma_{10} + \lambda_2)/\left\{\beta_{10}^2 + (\gamma_{10} + \lambda_2)^2\right\}^{1/2}
\end{aligned}.$$

Then the transformation $X = TY$, where $T = \begin{bmatrix} x_1 & x_1' \\ x_2 & x_2' \end{bmatrix}$ and $Y = \begin{bmatrix} Y_1 \\ Y_2 \end{bmatrix}$, converts Eq. (5.3.8) to the form

$$\frac{\mathrm{d}Y}{\mathrm{d}t} = [\Lambda + \varepsilon^2 B + \varepsilon f(t) P]Y \tag{5.3.9}$$

where

$$\begin{aligned}
\Lambda &= T^{-1}A^{(0)}T = \begin{bmatrix} \lambda_1 & 0 \\ 0 & \lambda_2 \end{bmatrix} \\
B &= T^{-1}B^{(0)}T = \frac{D}{i\beta_{10}\omega}\begin{bmatrix} b_{11} & b_{12} \\ b_{21} & b_{22} \end{bmatrix} \\
b_{11} &= -x_1(\gamma_{11}x_2' + \beta_{21}x_1') + x_2(\gamma_{21}x_1' - \beta_{11}x_2') \\
b_{12} &= -x_1'(\gamma_{11}x_2' + \beta_{21}x_1') + x_2'(\gamma_{21}x_1' - \beta_{11}x_2') \\
b_{21} &= x_1(\gamma_{11}x_2 + \beta_{21}x_1) + x_2(\beta_{11}x_2 - \gamma_{21}x_1) \\
b_{22} &= x_1'(\gamma_{11}x_2 + \beta_{21}x_1) + x_2'(\beta_{11}x_2 - \gamma_{21}x_1) \\
D &= \left[\left\{\beta_{10}^2 + (\gamma_{10} + \lambda_1)^2\right\}\left\{\beta_{10}^2 + (\gamma_{10} + \lambda_2)^2\right\}\right]^{1/2} \\
P &= T^{-1}P^{(0)}T = \frac{D}{i\beta_{10}\omega}\begin{bmatrix} p_{11} & p_{12} \\ p_{21} & p_{22} \end{bmatrix} \\
p_{11} &= -x_1(\gamma_{12}x_2' + \beta_{22}x_1') + x_2(\gamma_{22}x_1' - \beta_{12}x_2') \\
p_{12} &= -x_1'(\gamma_{12}x_2' + \beta_{22}x_1') + x_2'(\gamma_{22}x_1' - \beta_{12}x_2') \\
p_{21} &= x_1(\gamma_{12}x_2 + \beta_{22}x_1) + x_2(\beta_{12}x_2 - \gamma_{22}x_1) \\
p_{22} &= x_1'(\gamma_{12}x_2 + \beta_{22}x_1) + x_2'(\beta_{12}x_2 - \gamma_{22}x_1)
\end{aligned}$$

It is noted that, on account of the form chosen for T, we have $Y_1 = \overline{Y}_2$ since X is a real vector.

We seek for the solution of (5.3.9) of the form $Y = \mathrm{e}^{\Lambda t}Z(t)$, where

$$\mathrm{e}^{\Lambda t} = I + \Lambda t + \frac{\Lambda^2 t^2}{2!} + \frac{\Lambda^3 t^3}{3!} + \cdots; \; I = \begin{bmatrix} 1 & 0 \\ 0 & 1 \end{bmatrix}.$$

Then substituting it in (5.3.9), we obtain

$$\frac{\mathrm{d}Z}{\mathrm{d}t} = \varepsilon^2 B^{(1)}(t)Z + \varepsilon f(t)P^{(1)}(t)Z,$$

where

$$B^{(1)}(t) = e^{-\Lambda t} B e^{\Lambda t} = \frac{D}{i\beta_{10}\omega} \begin{bmatrix} b_{11} & b_{12}e^{-i\omega t} \\ b_{21}e^{i\omega t} & b_{22} \end{bmatrix}$$

$$P^{(1)}(t) = e^{-\Lambda t} P e^{\Lambda t} = \frac{D}{i\beta_{10}\omega} \begin{bmatrix} p_{11} & p_{12}e^{-i\omega t} \\ p_{21}e^{i\omega t} & p_{22} \end{bmatrix}$$

If the correlation time of the stochastic process $f(t)$ is sufficiently small, the process $Z(t)$ can be approximated over a time interval of order ε^{-2} by a homogeneous Markov vector that satisfies an Itô equation of the following form (Stratonovich 1968):

$$dZ = \varepsilon^2 m(Z)dt + \varepsilon\sigma(Z)d\xi$$

in which $\xi(t)$ is a 2-vector of independent Wiener process with unit variance and the coefficients $m(Z), \sigma(Z)$ are as follows:

$$m(Z) = M_t \left[B^{(1)}(t) + \int_{-\infty}^{\infty} P^{(1)}(t)P^{(1)}(t+\tau)\langle f(t)f(t+\tau)\rangle d\tau \right] Z$$

($\langle\cdot\rangle$ represents ensemble average over the stochastic process)

$$\left[\sigma(Z)\sigma'(Z)\right]_{rs} = \sum_{j,l=1}^{2} M_t \left[\int_{-\infty}^{\infty} p_{rj}(t)p_{sl}(t+\tau)\langle f(t)f(t+\tau)\rangle d\tau \right] Z_j Z_l$$

where M_t represents the averaging operator defined by

$$M_t = \lim_{T\to\infty} \frac{1}{T} \int_{t_0}^{t_0+\tau} (\cdot)dt$$

After some simplifications, we have

$$m(Z) = \begin{bmatrix} \left\{ \frac{D^2}{\omega^2\beta_{10}^2} \left(p_{11}^2 S(0) + p_{12}p_{21}S(\omega)\right) + i\frac{Db_{11}}{\beta_{10}\omega} \right\} Z_1 \\ \left\{ \frac{D^2}{\omega^2\beta_{10}^2} \left(p_{22}^2 S(0) + p_{12}p_{21}S(\omega)\right) + i\frac{Db_{22}}{\beta_{10}\omega} \right\} Z_2 \end{bmatrix}.$$

where

$$S(\omega) = \int_{-\infty}^{\infty} \langle f(t)f(t+\tau)\rangle \cos\omega\tau d\tau$$

represents the spectral density function of $f(t)$.

Applying the differential rule introduced by Itô (1951) to the function

$$\phi(Z) = Z_1 Z_2 = Z_1 \overline{Z_1} = |Z_1|^2 = |Z_2|^2$$

and taking the expectations, we get

$$\frac{d}{dt}\left\langle |Z_r|^2 \right\rangle = -\varepsilon^2 \Bigg[\frac{D}{\beta_{10}\omega}(b_{11} + b_{22})i + \frac{D^2}{\beta_{10}^2\omega^2}\left\{(p_{11}^2 + p_{11}p_{22} + p_{22}^2)S(0) + 3p_{12}p_{21}S(\omega)\right\} \Bigg] \left\langle |Z_r^2| \right\rangle,$$

where $r = 1, 2$.
After some simplifications, we have

$$\frac{d}{dt}\left\langle |Z_r|^2 \right\rangle = \varepsilon^2 \Omega \left\langle |Z_r^2| \right\rangle \quad (r = 1, 2)$$

where the expression for Ω is given in Theorem 2.1.
Therefore, the conditions for stability and instability in mean square are given by $\Omega \le 0$ and $\Omega > 0$, respectively. Following Kushner (1967) and Ariaratnam and Tam (1974), we conclude that $\Omega < 0$ guarantees asymptotic stability with probability one. □

Now, in the absence of self-interactions:

$$\Omega = \frac{1}{4\beta_{10}\beta_{20}}\left\{3(\beta_{10}\beta_{22} - \beta_{20}\beta_{12})^2 S(4\beta_{10}\beta_{20}) - (\beta_{10}\beta_{22} + \beta_{20}\beta_{12})^2 S(0)\right\}.$$

Therefore, if

$$3(\beta_{10}\beta_{22} - \beta_{20}\beta_{12})^2 S(4\beta_{10}\beta_{20}) < (\beta_{10}\beta_{22} + \beta_{20}\beta_{12})^2 S(0),$$

then the prey–predator system in the absence of self-interaction is asymptotically stable with probability one, whereas the deterministic system displays an oscillatory behaviour. The stability behaviour is, in fact, a consequence of stochastic parametric perturbations. It is also noted that in presence of self-interaction terms, the deterministic case is asymptotically stable, whereas for $\Omega > 0$, the stochastic case will be unstable. Thus, it would appear that instability may also be a consequence of stochastic parametric perturbations.

The stochastic process $f(t)$ may be chosen in different ways. Let us discuss the consequences of some important choices of $f(t)$.

I. Suppose $f(t)$ is the Ornstein–Uhlenbeck process or coloured noise whose mathematical expectation vanishes and correlation function is as follows:

$$\langle f(t)f(t+\tau)\rangle = \frac{\theta^2}{2\nu}\exp(-\nu|\tau|), \quad \nu > 0.$$

It decreases exponentially.
If the correlation time of this process $\tau_{\text{cor}} = \nu^{-1}$ is sufficiently small, i.e. ν is sufficiently large, then

$$S(\omega) = \int_{-\infty}^{\infty} \langle f(t)f(t+\tau)\rangle \cos\omega\tau \,\mathrm{d}\tau = \frac{\theta^2}{\omega^2+\nu^2}.$$

Let $\omega^2 = 2\nu^2$, which is sufficiently large. In this case:

$$\Omega = \frac{\theta^2}{\nu^2}\left\{(\gamma_{12}+\gamma_{22})^2 - (\gamma_{12}\gamma_{22}+\beta_{12}\beta_{22})\right\} - (\gamma_{11}+\gamma_{21})$$

from which we conclude that the system (5.3.6) is stable for

$$\gamma_{11}+\gamma_{21} \geq \frac{\theta^2}{\nu^2}\left\{(\gamma_{12}+\gamma_{22})^2 - (\gamma_{12}\gamma_{22}+\beta_{12}\beta_{22})\right\}$$

and unstable if

$$\gamma_{11}+\gamma_{21} < \frac{\theta^2}{\nu^2}\left\{(\gamma_{12}+\gamma_{22})^2 - (\gamma_{12}\gamma_{22}+\beta_{12}\beta_{22})\right\}.$$

In the absence of self-interaction, $\Omega = -\frac{\theta^2}{\nu^2}\beta_{12}\beta_{22} < 0$. Therefore in this case, the prey–predator system is asymptotically stable with probability one, whereas the deterministic system displays an oscillatory behaviour.

II. Let $f(t)$ be a random function with zero mean and correlation function (Wentzel and Ovcharov 1986):

$$\langle f(t)f(t+\tau)\rangle = (\nu^2-\theta^2)\mathrm{e}^{-\nu|\tau|}\left[\cosh\theta|\tau| - \frac{\nu}{\theta}\sinh\theta|\tau|\right], \quad (\nu > \theta > 0).$$

Then

$$\begin{aligned} S(\omega) &= \int_{-\infty}^{\infty} \langle f(t)f(t+\tau)\rangle \cos\omega\tau\,\mathrm{d}\tau \\ &= \frac{4\nu\omega^2(\nu^2-\theta^2)}{\{(\nu-\theta)^2+\omega^2\}\{(\nu+\theta)^2+\omega^2\}} \end{aligned}$$

and

$$\Omega = -(\gamma_{11} + \gamma_{21}) + 12\nu(\nu^2 - \theta^2)\{\beta_{10}\beta_{20}(\gamma_{22} - \gamma_{12})^2 + (\beta_{10}\beta_{22} - \beta_{12}\beta_{20})^2 \\ +(\gamma_{10} + \gamma_{20})(\gamma_{12} - \gamma_{22})(\beta_{10}\beta_{22} + \beta_{20}\beta_{12}) \\ +\beta_{12}\beta_{22}(\gamma_{10} - \gamma_{20})^2\} / \left[\{(\nu - \theta)^2 + \omega^2\}\{(\nu + \theta)^2 + \omega^2\}\right]$$

Therefore, the system (5.3.6) is stable or unstable according as

$$\{(\nu - \theta)^2 + \omega^2\}\{(\nu + \theta)^2 + \omega^2\}(\gamma_{11} + \gamma_{21}) \geq \text{or} < \Theta,$$

where

$$\Theta = 12\nu(\nu^2 - \theta^2)\{\beta_{10}\beta_{20}(\gamma_{22} - \gamma_{12})^2 + (\beta_{10}\beta_{22} - \beta_{12}\beta_{20})^2 \\ +(\gamma_{10} + \gamma_{20})(\gamma_{12} - \gamma_{22})(\beta_{10}\beta_{22} + \beta_{20}\beta_{12}) + \beta_{12}\beta_{22}(\gamma_{10} - \gamma_{20})^2\}.$$

In the absence of self-interaction,

$$\Omega = \frac{12\nu(\nu^2 - \theta^2)(\beta_{10}\beta_{22} - \beta_{12}\beta_{20})^2}{\left[\{(\nu - \theta)^2 + 4\beta_{10}\beta_{20}\}\{(\nu + \theta)^2 + 4\beta_{10}\beta_{20}\}\right]} > 0,$$

for $\beta_{10}\beta_{22} \neq \beta_{12}\beta_{20}$. Therefore, in this situation, the prey–predator system is unstable whereas the deterministic system exhibits an oscillatory behaviour.

5.4 Stochastic Gomatam Model of Predator–Prey Species: Non-equilibrium Fluctuation and Stability

Environmental fluctuations have a serious impact on real biological systems. Usually, such fluctuations do not strictly obey deterministic laws. It is observed that fluctuations will manifest themselves mainly as fluctuations in the intrinsic growth rate of the prey and in the mortality rate of the predator, since these are the main parameters subject to coupling of a predator–prey pair with its environment. In this article, we have studied the stability behaviour of the logarithm of the population size for the Gomatam system of interacting species by perturbing the prey's reproductive factor (intrinsic growth rate) and predator's mortality rate by coloured noises in order to take into account the effect of fluctuating environment. Here, we have used the spectral density technique developed by Pugachev (1963) which provides us with an expression of non-equilibrium fluctuation and associated criteria of stability or instability. It is shown that the system exhibits large fluctuations for high-amplitude random forces as a result of fluctuating environmental conditions. When the intraspecific interaction coefficients tend to zero along a straight line, this system exhibits abnormally large fluctuations with increasing time with a periodic background noise. We have also obtained a relation between the real part of the eigenvalues of the variational matrix (interaction matrix) in the deterministic environment and the stability of this system in the fluctuating (stochastic) environment. The biological relevance of the underlying model and the results obtained have been discussed.

5.4.1 *Modified Predator–Prey System (Gomatam Model): Basic Differential Equations*

In the Gomatam model (Gomatam 1974): (i) the Gompertz self-interaction term is generalized to interaction between two species, (ii) self-interaction (among the members of each species) is of the self-saturating Gompertz type. Therefore, the prey population size X and the predator population size Y will be considered within following framework:

$$\begin{aligned} \frac{dX}{dt} &= a_1 X - b_1 X \log Y - c_1 X \log X, \qquad X(0) > 0, \\ \frac{dY}{dt} &= -a_2 Y + b_2 Y \log X - c_2 Y \log Y, \qquad Y(0) > 0, \end{aligned} \tag{5.4.1}$$

where $a_j(> 0)$ represent the rate constants (intrinsic growth rate of the prey and the mortality rate of the predator respectively) in the absence of interaction, $b_j(> 0)$ are the binary interaction constants and $c_j(> 0)$ represent the coefficients of self-interaction. In a fluctuating environment, (5.4.1) becomes

$$\begin{aligned} \frac{dX}{dt} &= (a_1 + \xi_1(t))X - b_1 X \log Y - c_1 X \log X, \qquad X(0) > 0, \\ \frac{dY}{dt} &= -(a_2 + \xi_2(t))Y + b_2 Y \log X - c_2 Y \log Y, \qquad Y(0) > 0, \end{aligned} \tag{5.4.2}$$

where the perturbed terms $\xi_1(t)$ and $\xi_2(t)$ are independent coloured noises (or Ornstein–Uhlenbeck processes). It is experimentally recognized that coloured noises are more realistic noises than white noises (Horsthemke and Lefever 1984). The mathematical expectations and correlation functions of the processes $\xi_j(t)$ are respectively as follows:

$$\langle \xi_j(t) \rangle = 0 \text{ and } \langle \xi_j(t_1)\xi_j(t_2) \rangle = \epsilon_j \zeta_j \exp(-\zeta_j |t_1 - t_2|), \quad j = 1, 2, \tag{5.4.3}$$

where ϵ_j represents the intensity, $\zeta_j^{-1} > 0$ represents the correlation time of $\xi_j(t)$ and $\langle \cdot \rangle$ stands for the average over the ensemble of the stochastic process. The noises $\xi_j(t)$ are the solutions of the stochastic differential equations (Uhlenbeck and Ornstein 1954):

$$\frac{d\xi_j}{dt} = -\zeta_j \xi_j + \zeta_j \sqrt{2\epsilon_j} \frac{dw_j}{dt}, \tag{5.4.4}$$

where $\eta_j = \frac{dw_j}{dt}$ are independent standard zero mean Gaussian white noises characterized by the followings:

$$\langle \eta_j(t) \rangle = 0 \text{ and } \langle \eta_j(t_1)\eta_j(t_2) \rangle = \delta(t_1 - t_2), \quad j = 1, 2, \tag{5.4.5}$$

where $\delta(t)$ represents the Dirac delta function defined by

$$\begin{cases} \delta(t) = 0, \text{ for } t \neq 0, \\ \int_{-\infty}^{\infty} \delta(t)dt = 1. \end{cases}$$

Let $u = \log X$ and $v = \log Y$. Then the system of (5.4.2) can be rewritten as:

$$\begin{aligned} \frac{du}{dt} &= a_1 - c_1 u - b_1 v + \xi_1(t), \\ \frac{dv}{dt} &= -a_2 + b_2 u - c_2 v + \xi_2(t). \end{aligned} \tag{5.4.6}$$

In the absence of noise, system (5.4.6) has a unique interior equilibrium at the point:

$$(u^*, v^*) = \left(\frac{a_1c_2 + a_2b_1}{\Delta}, \frac{a_1b_2 - a_2c_1}{\Delta} \right), \quad \text{where } \Delta = c_1c_2 + b_1b_2.$$

It is observed that: (i) the limiting value of the prey population will be $u' = a_1/c_1$, in the absence of noise and predator, and (ii) the limiting value of the prey population will be $u'' = a_2/b_2$, in the absence of noise and intraspecific competition among the predator. It is immediately apparent that

$$u' > u'' \implies a_1b_2 - a_2c_1 > 0$$

makes v^* positive and hence interior equilibrium exists.

Substituting $p_1 = u - u^*, \quad p_2 = v - v^*$, system (5.4.6) boils down to:

$$\frac{dp_1}{dt} = -c_1p_1 - b_1p_2 + \xi_1(t), \tag{5.4.7a}$$

$$\frac{dp_2}{dt} = b_2p_1 - c_2p_2 + \xi_2(t). \tag{5.4.7b}$$

By eliminating p_2 from (5.4.7a) and p_1 from (5.4.7b):

$$\frac{d^2p_1}{dt^2} + (c_1 + c_2)\frac{dp_1}{dt} + (b_1b_2 + c_1c_2)p_1 = \Lambda_1(t), \tag{5.4.8}$$

where

$$\Lambda_1(t) = (c_2 - \zeta_1)\xi_1(t) + \zeta_1\sqrt{2\epsilon_1}\eta_1(t) - b_1\xi_2(t), \tag{5.4.9}$$

and

$$\frac{\mathrm{d}^2 p_2}{\mathrm{d}t^2} + (c_1 + c_2)\frac{\mathrm{d}p_2}{\mathrm{d}t} + (b_1 b_2 + c_1 c_2)p_2 = \Lambda_2(t), \tag{5.4.10}$$

where

$$\Lambda_2(t) = (c_1 - \zeta_2)\xi_2(t) + \zeta_2\sqrt{2\epsilon_2}\eta_2(t) + b_2\xi_1(t). \tag{5.4.11}$$

It is assumed that

$$|c_1 - c_2| < 2(b_1 b_2)^{1/2}.$$

5.4.2 Spectral Density Functions

The correlation function K_{Λ_1} of the driving force $\Lambda_1(t)$ is as follows:

$$\begin{aligned} K_{\Lambda_1}(t_1 - t_2) &= \langle(\Lambda_1(t_1) - \langle\Lambda_1(t_1)\rangle)(\Lambda_1(t_2) - \langle\Lambda_1(t_2)\rangle)\rangle \\ &= \langle\Lambda_1(t_1)\Lambda_1(t_2)\rangle \ \ (\because\ \langle\Lambda_1(t)\rangle = 0) \\ &= (c_2 - \zeta_1)^2\langle\xi_1(t_1)\xi_1(t_2)\rangle + 2\zeta_1^2\epsilon_1\langle\eta_1(t_1)\eta_1(t_2)\rangle + b_1^2\langle\xi_2(t_1)\xi_2(t_2)\rangle \\ &\quad + (c_2 - \zeta_1)\zeta_1\sqrt{2\epsilon_1}\{\langle\xi_1(t_1)\eta_1(t_2)\rangle + \langle\xi_1(t_2)\eta_1(t_1)\rangle\} \end{aligned} \tag{5.4.12}$$

$(\because \xi_1$ and ξ_2 are independent).

From (5.4.4):

$$\frac{\mathrm{d}\xi_1}{\mathrm{d}t} = -\zeta_1\xi_1(t) + \zeta_1\sqrt{2\epsilon_1}\eta_1(t),\ \ \langle\eta_1(t)\rangle = 0 \ \text{ and } \ \langle\eta_1(t_1)\eta_1(t_2)\rangle = \delta(t_1 - t_2).$$

Therefore, the correlation function K_Ξ of the random function $\Xi(t) = \dfrac{\mathrm{d}\xi_1(t)}{\mathrm{d}t}$ is as follows:

$$\begin{aligned} K_\Xi(\alpha) &= \langle\Xi(t_1)\Xi(t_2)\rangle \ \ (\because \langle\Xi(t)\rangle = 0) \\ &= \epsilon_1\zeta_1^3\mathrm{e}^{-\zeta_1|\alpha|} + 2\zeta_1^2\epsilon_1\delta(\alpha) - \zeta_1^2\sqrt{2\epsilon_1}\{\langle\xi_1(t_1)\eta_1(t_2)\rangle + \langle\xi_1(t_2)\eta_1(t_1)\rangle\}, \end{aligned} \tag{5.4.13}$$

where $\alpha = t_1 - t_2$. Again, from Wentzel and Ovcharov (1986), we have

$$K_\Xi(\alpha) = \epsilon_1\zeta_1^2\mathrm{e}^{-\zeta_1|\alpha|}\{2\delta(\alpha) - \zeta_1(\operatorname{sign}\alpha)^2\}, \tag{5.4.14}$$

where

$$\text{sign}\,\alpha = \begin{cases} 1, & \text{for } \alpha > 0, \\ 0, & \text{for } \alpha = 0, \\ -1, & \text{for } \alpha < 0. \end{cases}$$

Now, combining (5.4.13) and (5.4.14), we have

$$\begin{aligned} &\zeta_1\sqrt{2\epsilon_1}\{\langle\xi_1(t_1)\eta_1(t_2)\rangle + \langle\xi_1(t_2)\eta_1(t_1)\rangle\} \\ &\quad = \epsilon_1\zeta_1[\zeta_1 e^{-\zeta_1|\alpha|} + 2\delta(\alpha) + e^{-\zeta_1|\alpha|}\{\zeta_1(\text{sign}\,\alpha)^2\} - 2\delta(\alpha)\}]. \end{aligned} \tag{5.4.15}$$

Therefore, from (5.4.3), (5.4.5), (5.4.12) and (5.4.15), we have

$$\begin{aligned} K_{\Lambda_1}(\alpha) &= (c_2-\zeta_1)^2\epsilon_1\zeta_1 e^{-\zeta_1|\alpha|} + (c_2-\zeta_1)^2\epsilon_1\zeta_1[\zeta_1 e^{-\zeta_1|\alpha|} + 2\delta(\alpha) \\ &\quad + e^{-\zeta_1|\alpha|}\{\zeta_1(\text{sign}\,\alpha)^2 - 2\delta(\alpha)\}] + 2\zeta_1^2\epsilon_1\delta(\alpha) + b_1^2\epsilon_2\zeta_2 e^{-\zeta_2|\alpha|}. \end{aligned} \tag{5.4.16}$$

Similarly, the correlation function K_{Λ_2} of the driving force $\Lambda_2(t)$ is as follows:

$$\begin{aligned} K_{\Lambda_2}(\alpha) &= (c_1-\zeta_2)^2\epsilon_2\zeta_2 e^{-\zeta_1|\alpha|} + (c_1-\zeta_2)^2\epsilon_2\zeta_2[\zeta_2 e^{-\zeta_2|\alpha|} + 2\delta(\alpha) \\ &\quad + e^{-\zeta_2|\alpha|}\{\zeta_2(\text{sign}\,\alpha)^2 - 2\delta(\alpha)\}] + 2\zeta_2^2\epsilon_2\delta(\alpha) + b_2^2\epsilon_1\zeta_1 e^{-\zeta_1|\alpha|}. \end{aligned} \tag{5.4.17}$$

After some simplifications, the spectral density functions of $\Lambda_j(t)$ are given by:

$$\begin{aligned} S_{\Lambda_j}(\omega) &= \frac{1}{2\pi}\int_{-\infty}^{\infty} K_{\Lambda_j}(\alpha)e^{-i\omega\alpha}d\alpha, \qquad i=\sqrt{-1}, \\ &= \frac{1}{\pi}\left[\frac{\epsilon_j\zeta_j^2(\omega^2+c_{3-j}^2)}{\omega^2+\zeta_j^2} + \frac{2b_j^2\zeta_{3-j}^2\epsilon_{3-j}}{\omega^2+\zeta_{3-j}^2}\right], \quad j=1,2. \end{aligned} \tag{5.4.18}$$

5.4.3 Non-equilibrium Fluctuation and Stability

The mean square deviations of $p_j(t)$ $(j=1,2)$ at any arbitrary instant t are as follows (Pugachev 1963):

$$D_{p_j}(t) = \int_{-\infty}^{\infty} S_{\Lambda_j}(\omega)|y(t,\omega)|^2 d\omega, \quad j=1,2, \tag{5.4.19}$$

where $y(t,\omega)$ is the coordinate function which is given by the solution of equations (5.4.8) and (5.4.10) with $\Lambda_j(t)=e^{i\omega t}$, $p_j(t)=y(t,\omega)$ and $y(0,\omega)=0$.

$$\therefore\ y(t,\omega) = \frac{e^{i\omega t}}{\mu_0+\mu_1 i\omega-\omega^2} + \frac{(\lambda_2-i\omega)e^{\lambda_1 t}-(\lambda_1-i\omega)e^{\lambda_2 t}}{(\lambda_1-\lambda_2)(\mu_0+\mu_1 i\omega-\omega^2)}, \tag{5.4.20}$$

where $\lambda_1 = -\beta + i\gamma, \lambda_2 = -\beta - i\gamma, \beta = (c_1 + c_2)/2, \gamma = \{4b_1b_2 - (c_1 - c_2)^2\}^{1/2}/2$, $\mu_0 = b_1b_2 + c_1c_2$, and $\mu_1 = c_1 + c_2$.
Thus

$$D_{p_j}(t) = \frac{\epsilon_j\zeta_j^2}{\pi}\left\{(c_{3-j}^2 - \zeta_j^2)\int_{-\infty}^{\infty}\frac{|y(t,\omega)|^2}{\omega^2+\zeta_j^2}\mathrm{d}\omega + \int_{-\infty}^{\infty}|y(t,\omega)|^2\mathrm{d}\omega\right\}$$

$$+\frac{2b_j^2\zeta_{3-j}^2\epsilon_{3-j}}{\pi}\int_{-\infty}^{\infty}\frac{|y(t,\omega)|^2}{\omega^2+\zeta_{3-j}^2}\mathrm{d}\omega. \quad (5.4.21)$$

After some simplifications using (Gradshtyen and Ryzhik 1980):

$$D_{p_j}(t) = \epsilon_j\zeta_j^2(c_{3-j}^2 - \zeta_j^2)A_j(t) + \epsilon_j\zeta_j^2 B(t) + 2b_j^2\zeta_{3-j}^2\epsilon_{3-j}A_{3-j}(t), \quad (5.4.22)$$

where

$$\begin{aligned}
A_j(t) = \frac{1}{\mu_1}\Bigg[&\frac{\zeta_j+\mu_1}{\zeta_j\mu_0(\zeta_j^2+\mu_1\zeta_j+\mu_0)}\left\{1+\left(\frac{\mu_1}{2\gamma}\sin\gamma t+\cos\gamma t\right)^2\mathrm{e}^{-\mu_1 t}\right\}\\
&-2\left(\frac{\mu_1}{2\gamma}\sin\gamma t+\cos\gamma t\right)\mathrm{e}^{-(\mu_1/2)t}\{(\mu_0+\zeta_j^2)^2-\mu_1^2\zeta_j^2\}^{-1}\\
&\times\Bigg\{\frac{\mu_1}{\zeta_j}\mathrm{e}^{-\zeta_j t}-\frac{2\sqrt{\mu_0}}{\sqrt{4\mu_0-\mu_1^2}}\exp(-t\sqrt{\mu_0}\cos\phi)\sin(\phi-t\sqrt{\mu_0}\sin\phi)\\
&+\frac{2(2\mu_0-\mu_1^2+\zeta_j^2)}{\sqrt{\mu_0(4\mu_0-\mu_1^2)}}\exp(-t\sqrt{\mu_0}\cos\phi)\sin(\phi+t\sqrt{\mu_0}\sin\phi)\Bigg\}\\
&-\frac{2}{\gamma}\mathrm{e}^{-(\mu_1/2)t}(\mu_0-\mu_1\zeta_j+\zeta_j^2)^{-1}(\mu_0+\mu_1\zeta_j+\zeta_j^2)^{-1}\sin\gamma t\\
&\times\Bigg\{\mu_1\mathrm{e}^{-\zeta_j t}-\frac{2\mu_0}{\sqrt{4\mu_0-\mu_1^2}}\exp(-t\sqrt{\mu_0}\cos\phi)\sin(2\phi-t\sqrt{\mu_0}\sin\phi)\\
&+\frac{2(2\mu_0-\mu_1^2+\zeta_j^2)}{\sqrt{4\mu_0-\mu_1^2}}\exp(-t\sqrt{\mu_0}\cos\phi)\sin(t\sqrt{\mu_0}\sin\phi)\Bigg\}\\
&+\frac{\mathrm{e}^{\mu_1 t}\sin^2\gamma t}{\gamma^2(\zeta_j^2+\mu_1\zeta_j+\mu_0)}\Bigg],\quad j=1,2;
\end{aligned} \quad (5.4.23)$$

$$
\begin{aligned}
B(t) = & \frac{1}{\mu_0\mu_1}\left\{1+\left(\frac{\mu_1}{2\gamma}\sin\gamma t+\cos\gamma t\right)^2 \mathrm{e}^{-\mu_1 t}\right\} \\
& -2\left(\frac{\mu_1}{2\gamma}\sin\gamma t+\cos\gamma t\right)\mathrm{e}^{-(\mu_1/2)t} \\
& \times \exp(-t\sqrt{\mu_0}\cos\phi)\frac{\sin(\phi+t\sqrt{\mu_0}\sin\phi)}{\mu_0^{3/2}\sin 2\phi} \\
& -\frac{2}{\mu_0\gamma}\mathrm{e}^{-(\mu_1/2)t}\sin\gamma t\exp(-t\sqrt{\mu_0}\cos\phi) \\
& \times\frac{\sin(t\sqrt{\mu_0}\sin\phi)}{\sin 2\phi}+\frac{1}{\mu_1\gamma^2}\mathrm{e}^{-\mu_1 t}\sin^2\gamma t,
\end{aligned}
\tag{5.4.24}
$$

and

$$
\cos 2\phi = (\mu_1^2 - 2\mu_0)/2\mu_0, \quad |\phi| < \frac{\pi}{2}. \tag{5.4.25}
$$

$\therefore D_{p_j}(t) \to D_{p_j}(\infty)$ as $t \to \infty$ ($\because |\phi| < \pi/2;\ \mu_1 = c_1 + c_2 > 0$), where

$$
D_{p_j}(\infty) = \frac{\epsilon_j\zeta_j\{c_{3-j}^2(\zeta_j+\mu_1)+\mu_0\zeta_j\}}{\mu_0\mu_1(\zeta_j^2+\mu_1\zeta_j+\mu_0)} + \frac{2b_j^2\zeta_{3-j}\epsilon_{3-j}(\zeta_{3-j}+\mu_1)}{\mu_0\mu_1(\zeta_{3-j}^2+\mu_1\zeta_{3-j}+\mu_0)}, \quad j = 1, 2. \tag{5.4.26}
$$

It is evident that the convergence are rapid except when $\mu_1 = c_1 + c_2 (> 0)$ is very large. As correlation times of the coloured noises $\xi_j(t)$, $\nu_j^{\mathrm{cor}} = \zeta_j^{-1} \to 0+$, $\xi_j(t) \to \sqrt{2\epsilon_j}\eta_j(t)$, where $\eta_j(t)$ are the independent zero mean standard Gaussian white noises characterized by (5.4.5) (Horsthemke and Lefever 1984). In this situation, $D_{p_j}(\infty) \to \{\epsilon_j(c_{3-j}^2 + \mu_0) + 2b_j^2\epsilon_{3-j}\}/(\mu_0\mu_1)$ and so for high-amplitude random forces, $D_{p_j}(\infty)$ are large and the system exhibits large fluctuations. The equilibrium, which is stable in the absence of fluctuating environment, becomes unstable as a result of fluctuating environmental conditions. It agrees with Svirezhev and Logofet (1983).

In the deterministic environment: the eigenvalues of the variational matrix of the underlying system are $\lambda_{1,2} = -\beta \pm i\gamma$, where $\beta = (c_1 + c_2)/2 = \mu_1/2$. Therefore, μ_1 can be treated as the stability determining quantity and the deterministic stability criteria is fulfilled (since $\mu_1 = c_1 + c_2 > 0$). In the stochastic (fluctuating) environmental conditions: μ_1 can also be treated as the stability determining quantity, but in this situation it is no longer enough that $\mu_1 > 0$, because if $\epsilon_j \gg \mu_1$, population exhibit large fluctuations leading to rapid extinction. It is remembered that ϵ_j is the intensity of the environmental noise $\xi_j(t)$. When ϵ_j and μ_1 are commensurate: the system is likely to undergo significant fluctuations, even though they persist for long times. For the situations $\epsilon_j \ll \mu_1$: population fluctuations are relatively small, and so, the environment is effectively deterministic (non-fluctuating). These results agree with May (1973).

5.4.4 Special Situation

Assuming $c_2 = cc_1$ $(c > 0)$ and taking the limit as $c_1 \to 0+$ (after some simplifications using L'Hospital rule):

$$\lim_{c_1\to 0+} D_{p_j}(t) = \epsilon_j \zeta_j^2 \Bigg[\frac{t}{\mu_0' + \zeta_j^2} - \frac{\sin 2\sqrt{\mu_0'}t}{2\sqrt{\mu_0'}(\mu_0' + \zeta_j^2)} + \frac{2e^{-\zeta_j t}\zeta_j^2}{(\mu_0' + \zeta_j^2)^2} \left\{ \frac{1}{\zeta_j} \cos\sqrt{\mu_0'}t + \frac{1}{\sqrt{\mu_0'}} \sin\sqrt{\mu_0'}t \right\} - \frac{\zeta_j (1 + \cos^2\sqrt{\mu_0'}t)}{(\mu_0' + \zeta_j^2)^2} + \frac{\zeta_j^3 \sin^2\sqrt{\mu_0'}t}{\mu_0'(\mu_0' + \zeta_j^2)^2} \Bigg]$$
$$+ 2b_j^2 \epsilon_{3-j} \zeta_{3-j}^2 \Bigg[\frac{t}{\mu_0'(\mu_0' + \zeta_{3-j}^2)} - \frac{\sin 2\sqrt{\mu_0'}t}{2\mu_0'^{3/2}(\mu_0' + \zeta_{3-j}^2)} - 2(\mu_0' + \zeta_{3-j}^2)^{-2} e^{-\zeta_{3-j}t} \left\{ \frac{1}{\zeta_{3-j}} \cos\sqrt{\mu_0'}t + \frac{1}{\sqrt{\mu_0'}} \sin\sqrt{\mu_0'}t \right\}$$
$$- \frac{(1 + \cos^2\sqrt{\mu_0'}t)}{\zeta_{3-j}(\mu_0' + \zeta_{3-j}^2)^2} - \frac{\zeta_{3-j} \sin^2\sqrt{\mu_0'}t}{\mu_0'(\mu_0' + \zeta_{3-j}^2)^2} \Bigg], \quad j = 1, 2, \quad (5.4.27)$$

where $\mu_0' = b_1 b_2$.

It can be verified that

$$g_j\big|_{t=0} = \frac{dg_j}{dt}\bigg|_{t=0} = \frac{d^2 g_j}{dt^2}\bigg|_{t=0} = 0 \text{ and } \frac{d^3 g_j}{dt^3}\bigg|_{t=0} > 0,$$

where $g_j = \lim_{c_1\to 0+} D_{p_j}(t), \ j = 1, 2.$

Therefore, as $c_1 \to 0+$ along the straight line $c_2 = cc_1, \ c > 0$, the system of interacting (prey–predator) species in a random (fluctuating) environment exhibits abnormally large periodic fluctuations with increasing time. These results agree with Nicolis and Prigogine (1977), Gardiner (1983).

As correlation time of $\xi_j(t), \ \nu_j^{\text{cor}} = \zeta_j^{-1} \to 0+, \ \xi_j(t) \to \sqrt{2\epsilon_j}\eta_j(t)$, we have for this situation:

$$\lim_{\zeta_j\to\infty} \lim_{c_1\to 0+} D_{p_j}(t) \to \infty, \ \forall\, t > 0, \ \text{ where } c_2 = cc_1, \ c > 0,$$

leads the system always unstable, which implies that the system is never stable because there is no internal mechanism to reestablish the stability.

5.5 Damped Volterra–Lotka Prey–Predator System in a Rapidly Fluctuating Random Environment

The classical Volterra–Lotka equations are generally used to describe the time evolution of interacting prey–predator system. The introduction of intraspecific competition among the prey, resulting from the limited resources, makes the Volterra–Lotka model rough and the system is known as the *damped Volterra–Lotka system*. Rapidly, fluctuating environmental variations usually cause random variations in system parameters, in particular, in the natural growth coefficient of the prey and in the natural mortality of the predator. In this article, we have developed a general stochastic analysis of the behaviour of the damped Volterra–Lotka prey–predator system in a rapidly fluctuating random environment. The method is based on the technique of perturbation approximation of nonlinear-coupled stochastic differential equations.

5.5.1 *Damped Volterra–Lotka System: Basic Stochastic Differential Equations*

For damped Volterra–Lotka system, the prey population $x(t)$ and the predator population $y(t)$ satisfy the following deterministic equation:

$$\frac{\mathrm{d}}{\mathrm{d}t}X(t) = F(X(t)), \tag{5.5.1}$$

where

$$X(t) = \begin{bmatrix} x(t) \\ y(t) \end{bmatrix}, \quad F = \begin{bmatrix} x(t)(\alpha - kx(t) - \beta y(t)) \\ \\ y(t)(-m + \eta\beta x(t)) \end{bmatrix}, \text{ and } \alpha, m, k, \eta, \beta > 0.$$

For random environment, the parameters α and $-m$ will fluctuate randomly, and we then replace α and $-m$ by $\alpha + \theta_1(t)$ and $-m + \theta_2(t)$, respectively, where $\theta_1(t)$ and $\theta_2(t)$ are random fluctuating terms. We assume that these fluctuations are rapid, and we express this fact by writing $\theta(\tau) \equiv (\theta_1(\tau), \theta_2(\tau))$ where $\tau = t/\epsilon$ and $0 < \epsilon << 1$ is a small, non-random parameter.

In a rapidly fluctuating random environment, the stochastic modification of (5.5.1) is as follows:

$$\frac{\mathrm{d}}{\mathrm{d}t}X(\epsilon, t) = F(X(\epsilon, t), \theta(t/\epsilon)), \tag{5.5.2}$$

where

$$X(\epsilon, t) = \begin{bmatrix} x(\epsilon t) \\ y(\epsilon, t) \end{bmatrix}, \quad F(X(\epsilon, t), \theta(t/\epsilon)) = \begin{bmatrix} x(\epsilon, t)(\alpha + \theta_1(t/\epsilon) - kx(\epsilon, t) - \beta y(\epsilon, t)) \\ \\ y(\epsilon, t)(-m + \theta_2(t/\epsilon) + \eta\beta x(\epsilon, t)) \end{bmatrix},$$

$x(\epsilon, t), y(\epsilon, t)$ represent the prey and predator population, respectively, and $\alpha, m, k, \eta, \beta > 0;\ 0 < \epsilon << 1$.

The meaning of this is as follows: as the *natural* time t changes by a typical amount $\delta t,\ \theta(t/\epsilon)$ fluctuates considerably, since it experiences an elapsed time $\delta\tau = \dfrac{\delta t}{\epsilon}$ which is large when ϵ is small. We assume that the perturbed terms $\theta_1(\tau), \theta_2(\tau);\ \tau = t/\epsilon$, are coloured noises or Ornstein–Uhlenbeck processes. The mathematical expectations and correlation functions of these processes are given by

$$\begin{aligned} &\langle\theta_i(\tau)\rangle = 0,\ \langle\theta_i(\tau_1)\theta_i(\tau_2)\rangle = \frac{\sigma_i^2}{2\gamma_i}\exp(-\gamma_i|\tau_1 - \tau_2|), \\ &\langle\theta_1(\tau_1)\theta_2(\tau_2)\rangle = \sigma\exp(-|\tau_1 - \tau_2|)(1 + |\tau_1 - \tau_2|),\ (\gamma_i > 0,\ i = 1, 2), \end{aligned} \tag{5.5.3}$$

where $\langle\cdot\rangle$ represents the average over the ensemble of the stochastic process.

This is motivated by the fact that

$$\lim_{|\tau_1-\tau_2|\to\infty} \langle\theta_1(\tau_1)\theta_2(\tau_2)\rangle = 0 \Rightarrow \theta_1(\tau_1), \theta_2(\tau_2) \text{ tend to independent random processes.}$$

It is also noted that as $\sigma_i, \gamma_i \to \infty$ keeping $\dfrac{\sigma_i^2}{\gamma_i^2}$ = constant = D_i^2 (say), then $\theta_i(\tau) \to D_i\eta_i(\tau)$ where $\eta_i(\tau)$ are standard white noises, i.e. $\langle\eta_i(\tau)\rangle = 0,\ \langle\eta_i(\tau_1)\eta_i(\tau_2)\rangle = \delta(\tau_1 - \tau_2)$.

5.5.2 *Perturbation Approximation and Non-equilibrium Fluctuation*

We shall now use a two-term perturbation approximation to $X(\epsilon, t)$ (White 1977):

$$X(\epsilon, t) \sim X^0(t) + \sqrt{\epsilon}Y^0(t). \tag{5.5.4}$$

The first approximation

$$X^0(t) = \begin{bmatrix} x^0(t) \\ y^0(t) \end{bmatrix},$$

satisfies

$$\frac{\mathrm{d}}{\mathrm{d}t}X^0(t) = \overline{F}(X^0(t)), \tag{5.5.5}$$

where

$$\overline{F}(X^0(t)) = \lim_{T\to\infty} \frac{1}{T} \int_0^T \langle F(X^0(t),\ \theta(\tau))\rangle \mathrm{d}\tau$$

$$= \begin{bmatrix} x^0(t)(\alpha - kx^0(t) - \beta y^0(t)) \\ y^0(t)(-m + \eta\beta x^0(t)) \end{bmatrix}.$$

These are just the equations of the damped Volterra–Lotka system in a fixed environment. This system has a unique non-trivial equilibrium (both components of which are nonzero) at the point

$$\overline{X} = \begin{bmatrix} x^\star \\ y^\star \end{bmatrix},\ \text{where } x^\star = \frac{m}{\eta\beta},\ y^\star = \frac{\eta\alpha\beta - km}{\eta\beta^2} \tag{5.5.6}$$

It is immediately apparent that in the absence of predators, the limit value of prey population will be $x' = \frac{\alpha}{k}$. The realization of the obvious condition:

$$x^\star < x' \Rightarrow \eta\alpha\beta - km > 0 \tag{5.5.7}$$

which makes $y^\star$ positive, and hence, this equilibrium exists.

We assume that the system is at the initial time $t = 0$ at $\overline{X}$; therefore, we have $X^0(t) = \overline{X}$.

Here,

$$Y^0(t) = \begin{bmatrix} Y_1^0(t) \\ Y_2^0(t) \end{bmatrix}$$

is a Gaussian random process which satisfies the linear equation

$$\frac{\mathrm{d}}{\mathrm{d}t} Y^0(t) = CY^0(t) + W(t), \tag{5.5.8}$$

where

$$Y^0(0) = O,\ C = \frac{\partial \overline{F}}{\partial X}(\overline{X}) = \begin{bmatrix} -\frac{km}{\eta\beta} & -\frac{m}{\eta} \\ \frac{\eta\alpha\beta - km}{\beta} & 0 \end{bmatrix}, \tag{5.5.9}$$

and

$$\langle W(t)\rangle = O,\ \langle W(t)W^T(t')\rangle = A\delta(t - t'), \tag{5.5.10}$$

$$A = \lim_{T\to\infty} \frac{1}{T} \int_0^T \int_0^T \langle [F(\overline{X}, \theta(\tau_1)) - \langle F(\overline{X}, \theta(\tau_1))\rangle] [F(\overline{X}, \theta(\tau_2)) - \langle F(\overline{X}, \theta(\tau_2))\rangle]^T \rangle \mathrm{d}\tau_1 \mathrm{d}\tau_2$$

$$= \begin{bmatrix} A_{11} & A_{12} \\ A_{21} & A_{22} \end{bmatrix}, \tag{5.5.11}$$

$$A_{11} = \frac{m^2\sigma_1^2}{\eta^2\beta^2\gamma_1^2},\ A_{12} = A_{21} = \frac{4m(\eta\alpha\beta - km)\sigma}{\eta^2\beta^3},\ A_{22} = \frac{(\eta\alpha\beta - km)^2\sigma_2^2}{\eta^2\beta^4\gamma_2^2}. \tag{5.5.12}$$

Now the solution of (5.5.8) is given by

$$Y^0(t) = Y(t) \int_0^t Y^{-1}(s) W(s) \mathrm{d}s, \tag{5.5.13}$$

where $Y(t)$ satisfies the linear equation:

$$\frac{\mathrm{d}}{\mathrm{d}t} Y(t) = CY(t),\ Y(0) = I. \tag{5.5.14}$$

Therefore,

$$\langle Y^0(t)\rangle = O,\ \text{since } \langle W(s)\rangle = O. \tag{5.5.15}$$

The solution of (5.5.14) is given by

$$Y(t) = \begin{bmatrix} \frac{1}{\sqrt{\Delta}}\left(\lambda_1 e^{\lambda_1 t} - \lambda_2 e^{\lambda_2 t}\right) & -\frac{m}{\eta\sqrt{\Delta}}\left(e^{\lambda_1 t} - e^{\lambda_2 t}\right) \\ \frac{b}{\sqrt{\Delta}}\left(e^{\lambda_1 t} - e^{\lambda_2 t}\right) & -\frac{1}{\sqrt{\Delta}}\left(\lambda_2 e^{\lambda_1 t} - \lambda_1 e^{\lambda_2 t}\right) \end{bmatrix}, \tag{5.5.16}$$

where

$$\Delta = a^2 - \frac{4m}{\eta} b,\ a = \frac{km}{\eta\beta},\ b = \frac{\eta\alpha\beta - km}{\beta},\ \lambda_1 = \frac{-a + \sqrt{\Delta}}{2},\ \lambda_2 = \frac{-a - \sqrt{\Delta}}{2}. \tag{5.5.17}$$

Now, the expression of the strength of the fluctuation $D(t)$, the covariance at one instant of time, is given by

$$D(t) = \langle Y^0(t) Y^{0^T}(t) \rangle = Y(t) \left[\int_0^t Y^{-1}(s) A Y^{-1^T}(s) \mathrm{d}s \right] Y^T(t) \tag{5.5.18}$$

$$= \begin{bmatrix} D_{11}(t) & D_{12}(t) \\ D_{21}(t) & D_{22}(t) \end{bmatrix}, \text{ where}$$

$$D_{11}(t) = \frac{\mathrm{e}^{2\lambda_1 t}}{2\Delta b} \Phi(\lambda_1, \lambda_2) + \frac{\mathrm{e}^{2\lambda_2 t}}{2\Delta b} \Phi(\lambda_2, \lambda_1) + \frac{2m}{\Delta \eta} \mathrm{e}^{-at} \Psi + \frac{1}{2ab} \left\{ A_{11} b + A_{22} \frac{m}{\eta} \right\}, \tag{5.5.19}$$

$$D_{12}(t) = D_{21}(t) = \frac{\mathrm{e}^{2\lambda_1 t}}{2\Delta \lambda_1} \Phi(\lambda_1, \lambda_2) + \frac{\mathrm{e}^{2\lambda_2 t}}{2\Delta \lambda_2} \Phi(\lambda_2, \lambda_1) - \frac{\mathrm{e}^{-at}}{2\Delta b} \Theta - \frac{A_{22}}{2b}, \tag{5.5.20}$$

$$D_{22}(t) = \frac{\mathrm{e}^{2\lambda_1 t}}{2\Delta \lambda_1} Z(\lambda_2) + \frac{\mathrm{e}^{2\lambda_2 t}}{2\Delta \lambda_2} Z(\lambda_1) + \frac{2b}{\Delta} \mathrm{e}^{-at} \Psi + M, \tag{5.5.21}$$

$$\Phi(x, y) = A_{11} b x - 2 A_{12} x y + A_{22} \frac{m}{\eta} y, \quad \Psi = \frac{1}{a} \left\{ A_{11} b + A_{22} \frac{m}{\eta} \right\} + A_{12},$$

$$\Theta = 2 A_{11} b^2 + 2 A_{12} a b + A_{22} \left\{ a^2 + 2 \frac{m}{\eta} b \right\},$$

$$Z(x) = A_{11} b^2 - 2 A_{12} b x + A_{22} x^2,$$

$$M = \frac{A_{11} b \eta}{2am} + \frac{A_{22} \eta}{2amb} \left\{ a^2 + \frac{m}{\eta} b \right\} + \frac{A_{12}}{m} \eta. \tag{5.5.22}$$

Therefore, $D(t)$ converges exponentially to the limiting variance

$$D(\infty) = \begin{bmatrix} \frac{1}{2\eta\beta^2 k} \left\{ \frac{m\beta\sigma_1^2}{\gamma_1^2} + \frac{(\eta\alpha\beta - km)\sigma_2^2}{\eta\gamma_2^2} \right\} & \frac{(km - \eta\alpha\beta)\sigma_2^2}{2\eta^2\beta^3\gamma_2^2} \\ \frac{(km - \eta\alpha\beta)\sigma_2^2}{2\eta^2\beta^3\gamma_2^2} & \frac{(\eta\alpha\beta - km)}{\beta^2} \left\{ \frac{\sigma_1^2}{2k\gamma_1^2} + \frac{(k^2 m + \eta\beta(\eta\alpha\beta - km))\sigma_2^2}{2\eta^2\beta^2 km\gamma_2^2} + \frac{4\sigma}{\eta} \right\} \end{bmatrix}. \tag{5.5.23}$$

This convergence is rapid except when k is close to zero. In the case of high-amplitude fluctuations, $D_{11}(\infty)$ and $D_{22}(\infty)$ ars large. This indicates that for high-amplitude fluctuations, the system demonstrates statistical parametric shatter as a result of rapidly fluctuating environmental conditions, and the equilibrium, which is stable in the absence of these fluctuations, becomes unstable. This parametric shatter may occur not only for high-amplitude fluctuations but also for high fertility to the prey and for small η, β.

5.5.3 Special Case: Volterra–Lotka System

Now using (5.5.19) to (5.5.22), we have

$$\lim_{k\to 0} D_{11}(t) = \frac{1}{4}\sqrt{\frac{m}{\alpha}}\frac{1}{\eta^2\beta^2}\left[8\sigma\sqrt{m\alpha}\left\{\cos(2t\sqrt{m\alpha})-1\right\}+\left\{\frac{m\sigma_1^2}{\gamma_1^2}-\frac{\alpha\sigma_2^2}{\gamma_2^2}\right\}\sin(2t\sqrt{m\alpha})\right]$$
$$+\frac{m(m\sigma_1^2\gamma_2^2+\alpha\sigma_2^2\gamma_1^2)}{2\eta^2\beta^2\gamma_1^2\gamma_2^2}t \tag{5.5.24}$$

and

$$\lim_{k\to 0} D_{22}(t) = -\frac{1}{4}\sqrt{\frac{\alpha}{m}}\frac{1}{\beta^2}\left[8\sigma\sqrt{m\alpha}\left\{\cos(2t\sqrt{m\alpha})-1\right\}+\left\{\frac{m\sigma_1^2}{\gamma_1^2}-\frac{\alpha\sigma_2^2}{\gamma_2^2}\right\}\sin(2t\sqrt{m\alpha})\right]$$
$$+\frac{\alpha(m\sigma_1^2\gamma_2^2+\alpha\sigma_2^2\gamma_1^2)}{2\beta^2\gamma_1^2\gamma_2^2}t. \tag{5.5.25}$$

From the above results, we see that as $k \to 0+$ the damped Volterra–Lotka system tends to a classical Volterra–Lotka system in a random environment which demonstrates statistical parametric shatter with increasing time with a periodic background noise.

References

Ariaratnam, S.T., Tam, D.S.F.: Stability of weakly stochastic linear systems. In: Ariaratnam, S.T., Leipholz, H. (eds.) Symposium on Stochastic Problems in Mechanics (No. 10), pp. 183–192. University of Waterloo Press, Waterloo (1974)

Gardiner, C.W.: Handbook of Stochastic Methods. Springer-Verlag, Berlin (1983)

Gomatam, J.: A new model for interacting populations-P: two-species systems. Bull. Math. Biol. **36**, 347–353 (1974)

Gradshtyen, I.S., Ryzhik, I.M.: Table of Integrals: Series and Product. Academic Press, New York (1980)

Horsthemke, W., Lefever, R.: Noise-Induced Transitions. Springer-Verlag, New York (1984)
Itô, K.: On stochastic differential equations. Mem. Am. Math. Soc. **4**, 15 (1951)
Itô, K.: Differential equations determining Markov processes. Zenkoku Shijo Suguku Danwakai **1077**, 1352–1400 (1942)
Kushner, H.J.: Stochastic Stability and Control. Academic Press, New York (1967)
May, R.M.: Stability in randomly fluctuating versus deterministic environments. Am. Nat. **107**, 621–650 (1973)
Nicolis, G., Prigogine, I.: Self-Organization in Non-Equilibrium Systems. Wiley Interscience, New York (1977)
Pugachev, V.S.: Theory of Random Function with Application. Pergamon Press, Oxford (1963)
Risken, H.: The Fokker-Planck Equation. Springer-Verlag, Berlin (1984)
Schuss, Z.: Theory and Applications of Stochastic Differential Equations. Wiley, New York (1980)
Stratonovich, R.L.: Topics in the Theory of Random Noise, vol. I, II. Gordon and Breach, New York (1963)
Stratonovich, R.L.: Conditional Markov Processes and Their Application to the Theory of Optimal Control. American Elsevier, New York (1968)
Svirezhev, Yu.M., Logofet, D.O.: Stability of Biological Communities. MIR, Moscow (1983)
Uhlenbeck, G.E., Ornstein, L.S.: On the theory of Brownian motion. In: Wax, N. (ed.) Selected Papers on Noise and Stochastic Process. Dover, New York (1954)
Van Kampen, N.G.: Stochastic Processes in Physics and Chemistry. North-Holland Publishing Company, Oxford (1981)
Wentzel, E., Ovcharov, L.: Applied Problems in Probability Theory. MIR, Moscow (1986)
White, B.S.: The effects of a rapidly fluctuating random environment on systems of interacting species. SIAM J. Appl. Math. **32**, 666–693 (1977)

Chapter 6
Stochastic Analysis of a Demographic Model of Urbanization

6.1 Introduction

The study of long-term growth and development of urbanization is essential for optimum provision of adequate and efficient public services in the urban sector and for the formulation and evaluation of public policy. Hence, the need for developing quantitative models of urbanization for improved population predictions. It is expected that the analysis of such models would facilitate the comprehension of the underlying mechanism of the phenomenon of urbanization. There have been several attempts to formulate demographic models for urban growth; important among them are the quantitative models by Keyfitz (1980), Rogers (1968) and the United Nations (1980). These three demographic models are discussed in detail in an article by Ledent (1980).

The continuous time version of Rogers' demographic model (which was originally for discrete time) can be expressed by a pair of coupled linear differential equations (Ledent 1980). Denoting by $U(t)$ the urban population and by $R(t)$ the rural population, this model becomes

$$\begin{aligned} \frac{\mathrm{d}R}{\mathrm{d}t} &= (\alpha - m)R + \gamma U \\ \frac{\mathrm{d}U}{\mathrm{d}t} &= (\beta - \gamma)U + mR, \end{aligned} \tag{6.1}$$

where α is the natural growth rate of the rural population, β is that of the urban population, m is the in-migration rate per individual of rural population (from rural to urban sector) and γ is the out-migration rate per urbanite (from urban to rural area).

In this chapter, it is proposed to study the mean behaviour of the rural and urban population size for the demographic model described by system (6.1) with demographic parameters stochasticized in order to take account of the effect of fluctuating

© Springer Nature Singapore Pte Ltd. 2021

G. Samanta, *Deterministic, Stochastic and Thermodynamic Modelling of some Interacting Species*, Forum for Interdisciplinary Mathematics,
https://doi.org/10.1007/978-981-16-6312-3_6

environment. If we consider a physical situation in which the random environmental fluctuations do not occur with great rapidity in comparison with the time scale of population growth, it is natural to assume that the stochastic parameters have finite correlation time.

6.2 Modified Stochastic Demographic Model

The modified stochastic version of system (6.1) to be considered is

$$\begin{aligned}\frac{dX_1(t)}{dt} &= \beta_1(t)X_2(t) + \gamma_1(t)X_1(t)\\ \frac{dX_2(t)}{dt} &= \beta_2(t)X_1(t) + \gamma_2(t)X_2(t),\end{aligned} \tag{6.2}$$

where $X_1(t) = R(t),\ X_2(t) = U(t),\ \beta_1(t) = \gamma(t),\ \beta_2(t) = m(t),\ \gamma_1(t) = \alpha(t) - m(t)$ and $\gamma_2(t) = \beta(t) - \gamma(t)$. The stochastic parameters $\beta_i(t)$ and $\gamma_i(t)$ can be expressed as follows:

$$\begin{aligned}\beta_i(t) &= \beta_{i0}[1 + \epsilon_i \Delta(t)],\ i = 1, 2\\ \gamma_i(t) &= \gamma_{i0}[1 + \epsilon_{i+2}\Delta(t)],\ i = 1, 2\end{aligned} \tag{6.3}$$

where β_{i0} and γ_{i0} are constants and ϵ_i are also constants but positive. $\Delta(t)$ is assumed to be a dichotomic Markov process (DMP) (Brissaud and Frisch 1974) defined as a two-valued stepwise constant Markov process with equiprobable values $+1$ and -1 and transition probabilities (for $t_2 > t_1$):

$$\begin{aligned}&P\{\Delta(t_2) = +1|\Delta(t_1) = +1\} = P\{\Delta(t_2) = -1|\Delta(t_1) = -1\}\\ &= \frac{1}{2}[1 + e^{-\nu(t_2-t_1)}],\\ &P\{\Delta(t_2) = +1|\Delta(t_1) = -1\} = P\{\Delta(t_2) = -1|\Delta(t_1) = +1\}\\ &= \frac{1}{2}[1 - e^{-\nu(t_2-t_1)}],\end{aligned} \tag{6.4}$$

where $\frac{1}{\nu}$ is the correlation time.

For a DMP, we have

$$E[\Delta(t)] = 0,\ E[\Delta^2] = 1,\ E[\Delta(t_1)\Delta(t_2)] = e^{-\nu|t_1-t_2|}. \tag{6.5}$$

Also, the closure property holds exactly, that is, if $\Psi\{\Delta(\cdot)\}$ is a functional of the process involving only times prior to t_1, then

$$E[\Delta(t_1)\Delta(t_2)\Psi\{\Delta(\cdot)\}] = E[\Delta(t_1)\Delta(t_2)]E[\Psi\{\Delta(\cdot)\}]. \tag{6.6}$$

6.3 Averages of the Populations

System (6.2) can be written in matrix form:

$$\frac{\mathrm{d}X(t)}{\mathrm{d}t} = AX(t) + \Delta(t)BX(t), \tag{6.7a}$$

where

$$X(t) = \begin{bmatrix} X_1(t) \\ X_2(t) \end{bmatrix}, \ A = \begin{bmatrix} \gamma_{10} & \beta_{10} \\ \beta_{20} & \gamma_{20} \end{bmatrix}, \ B = \begin{bmatrix} \epsilon_3\gamma_{10} & \epsilon_1\beta_{10} \\ \epsilon_2\beta_{20} & \epsilon_4\gamma_{20} \end{bmatrix}. \tag{6.7b}$$

This has the solution

$$X(t) = \mathrm{e}^{At}X(0) + \int_0^t \mathrm{e}^{A(t-t')}\Delta(t')BX(t')\mathrm{d}t'. \tag{6.8}$$

Iterating,

$$X(t) = \mathrm{e}^{At}X(0) + \int_0^t \mathrm{d}t'\mathrm{e}^{A(t-t')}\Delta(t')B\left\{\mathrm{e}^{At'}X(0) + \int_0^{t'} \mathrm{e}^{A(t'-t'')}\Delta(t'')BX(t'')\mathrm{d}t''\right\}.$$

Taking the ensemble averages and using $E[\Delta(t)] = 0$, we get

$$E[X(t)] = \mathrm{e}^{At}X(0) + \int_0^t \mathrm{d}t'\mathrm{e}^{A(t-t')}B\left\{\int_0^{t'} E[\Delta(t')\Delta(t'')\mathrm{e}^{A(t'-t'')}BX(t'')]dt''\right\}. \tag{6.9}$$

$$\therefore E[X(t)] = e^{At}X(0) + \int_0^t e^{A(t-t')}B\Phi(t')dt';\ \Phi(t') = \int_0^{t'} e^{(A-\nu I)(t'-t'')}BE[X(t'')]dt''. \tag{6.10}$$

Taking Laplace transforms:

$$E[\overline{X}(s)] = [sI - A]^{-1}[X(0) + B\overline{\Phi}(s)] \text{ and } \overline{\Phi}(s) = [(s+\nu)I - A]^{-1}BE[\overline{X}(s)]$$
$$\Rightarrow \{sI - A - B[(s+\nu)I - A]^{-1}B\}E[\overline{X}(s)] = X(0). \tag{6.11}$$

Now writing $D = (s+\nu-\gamma_{10})(s+\nu-\gamma_{20}) - \beta_{10}\beta_{20}$, it is seen that

$$[(s+\nu)I - A]^{-1} = \frac{1}{D}\begin{bmatrix} s+\nu-\gamma_{20} & \beta_{10} \\ \beta_{20} & s+\nu-\gamma_{10} \end{bmatrix}.$$

Using (6.5):

$$B[(s+\nu)I - A]^{-1}B = \frac{1}{D}\begin{bmatrix} \mu_{11} & \mu_{12} \\ \mu_{21} & \mu_{22} \end{bmatrix}, \text{ where}$$

$$\mu_{11} = \epsilon_3^2\gamma_{10}^2(s+\nu-\gamma_{20}) + \epsilon_1\epsilon_3\beta_{10}\beta_{20}\gamma_{10} + \epsilon_2\epsilon_3\beta_{10}\beta_{20}\gamma_{10} + \epsilon_1\epsilon_2\beta_{10}\beta_{20}(s+\nu-\gamma_{10}),$$

$$\mu_{12} = \epsilon_1\epsilon_3\beta_{10}\gamma_{10}(s+\nu-\gamma_{20}) + \epsilon_1^2\beta_{10}^2\beta_{20} + \epsilon_3\epsilon_4\beta_{10}\gamma_{10}\gamma_{20} + \epsilon_1\epsilon_4\beta_{10}\gamma_{20}(s+\nu-\gamma_{10}),$$

$$\mu_{21} = \epsilon_2\epsilon_3\beta_{20}\gamma_{10}(s+\nu-\gamma_{20}) + \epsilon_3\epsilon_4\beta_{20}\gamma_{10}\gamma_{20} + \epsilon_2^2\beta_{10}\beta_{20}^2 + \epsilon_2\epsilon_4\beta_{20}\gamma_{20}(s+\nu-\gamma_{10}),$$

$$\mu_{22} = \epsilon_1\epsilon_2\beta_{10}\beta_{20}(s+\nu-\gamma_{20}) + \epsilon_1\epsilon_4\beta_{10}\beta_{20}\gamma_{20} + \epsilon_2\epsilon_4\beta_{10}\beta_{20}\gamma_{20} + \epsilon_4^2\gamma_{20}^2(s+\nu-\gamma_{10}).$$

$$\text{Writing}: \ F = \left(s-\gamma_{10}-\frac{\mu_{11}}{D}\right)\left(s-\gamma_{20}-\frac{\mu_{22}}{D}\right) - \left(\beta_{10}+\frac{\mu_{12}}{D}\right)\left(\beta_{20}+\frac{\mu_{21}}{D}\right),$$

we find that

$$\{sI - A - B[(s+\nu)I - A]^{-1}B\}^{-1} = F^{-1}\begin{bmatrix} s-\gamma_{20}-\frac{\mu_{22}}{D} & \beta_{10}+\frac{\mu_{12}}{D} \\ \beta_{20}+\frac{\mu_{21}}{D} & s-\gamma_{10}-\frac{\mu_{11}}{D} \end{bmatrix}. \tag{6.12}$$

From (6.11) and (6.12), we have

$$E[\overline{X_1}(s)] = F^{-1}\left\{\left(s - \gamma_{20} - \frac{\mu_{22}}{D}\right)X_1(0) + \left(\beta_{10} + \frac{\mu_{12}}{D}\right)X_2(0)\right\},$$
$$E[\overline{X_2}(s)] = F^{-1}\left\{\left(s - \gamma_{10} - \frac{\mu_{11}}{D}\right)X_2(0) + \left(\beta_{20} + \frac{\mu_{21}}{D}\right)X_1(0)\right\}. \tag{6.13}$$

$E[X_1(t)]$ and $E[X_2(t)]$ are, respectively, the inverse Laplace transform of (6.13), and their evaluation is laborious. System (6.13) gives the values of $E[\overline{X_1}(s)]$ and $E[\overline{X_2}(s)]$ for different magnitudes of stochastic parametric perturbations.

6.4 Discussion of the Stability of a Delta-Correlated Stochastic Process

Here, we consider a physical situation in which the environmental changes occur with great rapidity as compared to the time scale of population growth, and consequently, we take the random fluctuations due to them to be delta-correlated and the corresponding results can be obtained in the limit of infinitesimally short correlation time. To this end we let $\epsilon_i, \nu \to \infty$ such that $\frac{\epsilon_i^2}{\nu} \to c_i,\ i = 1, 2, 3, 4$. Then the autocorrelatlon function $E[\epsilon_i\Delta(t_1)\epsilon_i\Delta(t_1)] \to 2c_i\delta(t_1 - t_2)$.
Also,

$$\frac{\mu_{11}}{D} \to c_3\gamma_{10}^2 + \beta_{10}\beta_{20}\sqrt{c_1c_2},$$

$$\frac{\mu_{12}}{D} \to \beta_{10}\sqrt{c_1}[\gamma_{10}\sqrt{c_3} + \gamma_{20}\sqrt{c_4}],$$

$$\frac{\mu_{21}}{D} \to \beta_{20}\sqrt{c_2}[\gamma_{10}\sqrt{c_3} + \gamma_{20}\sqrt{c_4}],$$

$$\frac{\mu_{22}}{D} \to c_4\gamma_{20}^2 + \beta_{10}\beta_{20}\sqrt{c_1c_2}.$$

Thus for a delta-correlated process:

$$E[\overline{X_1}(s)] = G^{-1}\left\{\left(s - \gamma_{20} - c_4\gamma_{20}^2 - \beta_{10}\beta_{20}\sqrt{c_1c_2}\right)X_1(0)\right.$$
$$\left. + \beta_{10}\left(1 + \gamma_{10}\sqrt{c_1c_3} + \gamma_{20}\sqrt{c_1c_4}\right)X_2(0)\right\},$$
$$E[\overline{X_2}(s)] = G^{-1}\left\{\beta_{20}\left(1 + \gamma_{10}\sqrt{c_2c_3} + \gamma_{20}\sqrt{c_2c_4}\right)X_1(0)\right.$$
$$\left. + \left(s - \gamma_{10} - c_3\gamma_{10}^2 - \beta_{10}\beta_{20}\sqrt{c_1c_2}\right)X_2(0)\right\}, \tag{6.14}$$

where

$$G = (s - \gamma_{10} - c_3\gamma_{10}^2 - \beta_{10}\beta_{20}\sqrt{c_1c_2})(s - \gamma_{20} - c_4\gamma_{20}^2 - \beta_{10}\beta_{20}\sqrt{c_1c_2})$$
$$-\beta_{10}\beta_{20}(1 + \gamma_{10}\sqrt{c_1c_3} + \gamma_{20}\sqrt{c_1c_4})(1 + \gamma_{10}\sqrt{c_2c_3} + \gamma_{20}\sqrt{c_2c_4}).$$

(6.14) can be written as:

$$E[\overline{X_1}(s)] = \frac{(s+b)X_1(0) + cX_2(0)}{(s+a)(s+b)+d},$$
$$E[\overline{X_2}(s)] = \frac{(s+b)X_2(0) + c'X_1(0)}{(s+a)(s+b)+d}, \tag{6.15}$$

where

$$a = -(\gamma_{10} + c_3\gamma_{10}^2 + \beta_{10}\beta_{20}\sqrt{c_1c_2});\ b = -(\gamma_{20} + c_4\gamma_{20}^2 + \beta_{10}\beta_{20}\sqrt{c_1c_2});$$
$$c = \beta_{10}(1 + \gamma_{10}\sqrt{c_1c_3} + \gamma_{20}\sqrt{c_1c_4});\ c' = \beta_{20}(1 + \gamma_{10}\sqrt{c_2c_3} + \gamma_{20}\sqrt{c_2c_4});\ d = -cc'.$$

Let

$$\alpha, \beta = -\frac{1}{2}\left\{-(a+b) \pm \sqrt{(a-b)^2 - 4d}\right\}.$$

Taking the inverse Laplace transform of (6.15):

$$E[X_1(t)] = \frac{1}{\alpha - \beta}[\{(\alpha + b)X_1(0) + cX_2(0)\}e^{\alpha t} - \{(\beta + b)X_1(0) + cX_2(0)\}e^{\beta t}],$$
$$E[X_2(t)] = \frac{1}{\alpha - \beta}[\{(\alpha + a)X_2(0) + c'X_1(0)\}e^{\alpha t} - \{(\beta + a)X_2(0) + c'X_1(0)\}e^{\beta t}]. \tag{6.16}$$

Let us assume that the stochastic perturbation terms are the same. Hence $c_1 = c_2 = c_3 = c_4$. Now,

$$K^2 = (a-b)^2 - 4d = \{1 + c_1(\gamma_{20} + \gamma_{10})\}^2\{(\gamma_{20} - \gamma_{10})^2 + 4\beta_{10}\beta_{20}\} > 0. \tag{6.17}$$

For asymptotically large times, $E[X_1(t)]$ and $E[X_2(t)]$ are unstable for:

$$-(a+b) + K > 0 \Rightarrow \gamma_{10} + \gamma_{20} > -c_1(\gamma_{10}^2 + \gamma_{20}^2 + 2\beta_{10}\beta_{20}) - K. \tag{6.18}$$

If $\gamma_{10} + \gamma_{20} < -c_1(\gamma_{10}^2 + \gamma_{20}^2 + 2\beta_{10}\beta_{20}) - K$, then $E[X_1(t)]$ and $E[X_2(t)]$ will tend to zero, i.e. they will extinct after a long time span.

For

$$\gamma_{10} + \gamma_{20} = -c_1(\gamma_{10}^2 + \gamma_{20}^2 + 2\beta_{10}\beta_{20}) - K, \tag{6.19}$$

$$\begin{aligned} \lim_{t\to\infty} E[X_1(t)] &= \frac{1}{K}\{(\alpha + b)X_1(0) + cX_2(0)\}, \\ \lim_{t\to\infty} E[X_2(t)] &= \frac{1}{K}\{(\alpha + a)X_2(0) + c'X_1(0)\}. \end{aligned} \tag{6.20}$$

In this case, we have

$$\begin{aligned} \lim_{t\to\infty} \frac{E[X_1(t)]}{E[X_1(t)] + E[X_2(t)]} &= \frac{(\alpha + b)X_1(0) + cX_2(0)}{\Omega}, \\ \lim_{t\to\infty} \frac{E[X_2(t)]}{E[X_1(t)] + E[X_2(t)]} &= \frac{(\alpha + a)X_2(0) + c'X_1(0)}{\Omega}, \end{aligned} \tag{6.21}$$

where $\Omega = (\alpha + a + c)X_2(0) + (\alpha + b + c')X_1(0)$. After some simple calculations, we have

$$\begin{aligned} \alpha + a &= \{1 + c_1(\gamma_{20} + \gamma_{10})\}\left[\frac{1}{2}(\gamma_{20} - \gamma_{10}) + \left\{\frac{1}{4}(\gamma_{20} - \gamma_{10})^2 + \beta_{10}\beta_{20}\right\}^{1/2}\right] \\ \alpha + b &= \{1 + c_1(\gamma_{20} + \gamma_{10})\}\left[\frac{1}{2}(\gamma_{10} - \gamma_{20}) + \left\{\frac{1}{4}(\gamma_{20} - \gamma_{10})^2 + \beta_{10}\beta_{20}\right\}^{1/2}\right]. \end{aligned} \tag{6.22}$$

If $\gamma_{20} > \gamma_{10}, \beta_{20} \geq \beta_{10}$ or $\gamma_{20} \geq \gamma_{10}, \beta_{20} > \beta_{10}$, then $\alpha + a > c$ and $\alpha + b < c'$, therefore

$$\frac{(\alpha + b)X_1(0) + cX_2(0)}{\Omega} < \frac{(\alpha + a)X_2(0) + c'X_1(0)}{\Omega}.$$

Hence, it is concluded that if mean value of $(\beta - \gamma) >$ mean value of $(\alpha - m)$ and mean value of $m \geq$ mean value of γ or mean value of $(\beta - \gamma) \geq$ mean value of $(\alpha - m)$ and mean value of $m >$ mean value of γ, then we have a higher level of urbanization after a long time span.

If mean value of $(\beta - \gamma) =$ mean value of $(\alpha - m)$ and mean value of $m =$ mean value of γ, then the level of urban population and the level of rural population will be equal after a long time span.

It may also be noted that for $\beta_{10} = 0$ (no migration from urban areas), Roger's model reduces to Keyfitz's model. For this model, if the condition (6.19) holds good and mean value of $(\beta - \gamma) \geq$ mean value of $(\alpha - m)$, then

$$\lim_{t\to\infty} \frac{E[X_1(t)]}{E[X_1(t)]+E[X_2(t)]} = 0,$$
$$\lim_{t\to\infty} \frac{E[X_2(t)]}{E[X_1(t)]+E[X_2(t)]} = 1. \tag{6.23}$$

Therefore, in this case, the entire population will eventually be urbanized.

6.5 Discussion

In this chapter, an asymptotic analysis of a continuous stochastic version of Roger's demographic model is carried out. It is seen that under some conditions, this model holds up high level of urbanization. From this continuous stochastic version of Roger's demographic model, we have also deduced a continuous stochastic version of Keyfitz's demographic model; here, it is seen that under some conditions, the entire population will eventually be urbanized.

References

Brissaud, A., Frisch, U.: Solving linear stochastic differential equations. J. Math. Phys. **15**, 524–534 (1974)

Keyfitz, N.: Do cities grow by natural increase or migration? Geogr. Anal. **12**, 142–156 (1980)

Ledent, J.: Comparative dynamics of three demographic models of urbanization, IIASA Report, RR-80-1, 1980

Rogers, A.: Matrix Analysis of Interregional Populations Growth and Distributions. University of California Press, California (1968)

United Nations Population Division, Patterns of Urban and Rural Population Growth, New York, 1980

Chapter 7
Non-equilibrium Thermodynamics of Interacting Species

7.1 Introduction

Based on the Lotka–Volterra's pioneering work, the application of modern methods of analysis has permitted the presentation of a large number of mathematical models dealing with the trophic interactions among populations. The standard definition of ecosystem is a community of organisms and their physical environment interacting as an ecological unit. An ecosystem can be considered as a physical entity: a dynamical system consisting of biological entity, typically a regional biota (or community) together with its environment subject to physical laws. An ecosystem consisting of an innumerable number of interacting components in relation to its environment possesses a great deal of complexity. Thermodynamics, in particular non-equilibrium thermodynamics, appears to offer ecology a general framework for the development of a consistent theoretical model of complex ecosystems and provides a very powerful tool in establishing the physical description of ecosystems.

7.2 Interacting Species: Thermodynamic Model and Entropy-Production

An ecosystem, from the thermodynamic point of view, can be considered as an open system exchanging energy and nutrient biomass with the environment. For a consistent thermodynamic model of the ecosystem, let us make the following assumptions (Nicolis and Prigogine 1977; Chakrabarti et al. 1995):

- Ecosystem is a regional biota of a fixed volume consisting of n interacting species of population sizes N_i $(i = 1, 2, \ldots, n)$ embedded in an environment which is in a thermodynamic equilibrium at a temperature T.

© Springer Nature Singapore Pte Ltd. 2021

G. Samanta, *Deterministic, Stochastic and Thermodynamic Modelling of some Interacting Species*, Forum for Interdisciplinary Mathematics,
https://doi.org/10.1007/978-981-16-6312-3_7

- Ecosystem exchanges energy, nutrient and biomass with the environment or external world through a separating surface Σ (say). In general, the composition prevailing in the external world are not the same or identical with those inside Σ.
- The time evolution of the i-th species population N_i is assumed to be governed by the system of ODEs:

$$\frac{\mathrm{d}N_i}{\mathrm{d}t} = f_i(N_1, N_2, \ldots, N_n),\ i = 1, 2, \ldots, n. \tag{7.2.1}$$

- The internal energy U of the ecosystem is additive with respect to the population sizes of the species:

$$U = \sum_{i=1}^{n} N_i u_i, \tag{7.2.2}$$

where u_i denotes the internal energy per individual of the i-th species.

Entropy: The second law of thermodynamics postulates the existence of a function of state called *entropy* (from the Greek word meaning *evolution*) which possesses the following properties:

1. Entropy of a system is an *extensive* property. If a system consists of several parts, the total entropy is equal to the sum of the entropies of each part.
2. Change of entropy $\mathrm{d}S$ can be split into two parts: $\mathrm{d}S = \mathrm{d}S_{(i)} + \mathrm{d}S_{(e)}$, where $\mathrm{d}S_{(e)}$ represents the flow of entropy due to interactions with the exterior, and $\mathrm{d}S_{(i)}$ denotes the contribution due to changes inside the system. The entropy increase $\mathrm{d}S_{(i)}$ due to changes inside the system is always non-negative. It is zero when the system undergoes reversible changes only but it is positive if the system is subject to irreversible changes as well.

The evolutionary ecosystem is assumed to be open, and so, we can decompose the entropy S (which is a measure of disorder or uncertainty) of the ecosystem into two parts:

$$\mathrm{d}S = \mathrm{d}S_{(i)} + \mathrm{d}S_{(e)}, \tag{7.2.3}$$

where the suffix "(i)" represents the internal part and "(e)" denotes the external part, i.e. the contribution due to the exchange with its environment. According to the second law of thermodynamics, the entropy-productions $\mathrm{d}S_{(i)}$ for irreversible processes is always non-negative, i.e. $\mathrm{d}S_{(i)} > 0$. Thus, in an open system, it is possible to decrease the entropy of the system if $-\mathrm{d}S_{(e)} > \mathrm{d}S_{(i)} \Rightarrow \mathrm{d}S < 0$. Such systems termed as *systems of dissipative structures* maintaining their form or structure by continuous dissipation or consumption of energy. Biological structures can only originate in a dissipative medium and be maintained by a continuous supply of energy (Prigogine 1967). Non-living systems (like clouds) and living system (from organisms to ecosystems) are dependent on outside energy fluxes to maintain their organization.

Let us turn to find the expression of the rate of entropy-production. Gibbs relation for open system gives

$$dS = \frac{1}{T}dU + \frac{p}{T}dV - \sum_{i=1}^{n} \frac{\Delta\mu_i}{T} dN_i, \tag{7.2.4}$$

where p is the pressure, V is the volume occupied by the system, μ_i is the chemical potential of the i-th component or species. If we assume the ecosystem as a regional biota (community) of fixed volume, the second term of the right-hand side (7.2.4) vanishes. The relation (7.2.4) then shows that for the open system model of the ecosystem, the rate of entropy-production can be written in the form (using (7.2.2)):

$$\frac{dS}{dt} = \sum_{i=1}^{n} N_i s_i + \frac{d\overline{S}}{dt}, \tag{7.2.5}$$

where $s_i = \frac{1}{T}\frac{du_i}{dt}$ (by (7.2.2)) is the entropy-production rate per unit individual of the i-th species. The second term of the fight-hand side of (7.2.5) is connected with the rate of change of population size N_i:

$$\frac{d\overline{S}}{dt} = -\sum_{i=1}^{n} \frac{\Delta\mu_i}{T}\frac{dN_i}{dt}, \tag{7.2.6}$$

where μ_i is the chemical potential of the i-th component or species. In the above, we have assumed that the temperature remains constant during the evolutionary process. Let us define $\overline{S}$ as the *evolutionary entropy*. The expression (7.2.6) shows the rate of evolutionary entropy-production is governed by the rate of biomass production, i.e. by the dynamical equation of evolution (7.2.1). This is consistent with our model where the energy is assumed to be stored exclusively within the organism, where every changes or transfer of energy is determined by the ecological evolution equations (7.2.1).

Let us now, in particular, consider an ecosystem governed by the generalized Lotka–Volterra equations of evolution (Svirezhev and Logofet 1978):

$$\frac{dN_i}{dt} = \left\{ a_i - \sum_{j=1}^{n} b_{ij} N_j \right\} N_i, \; i = 1, 2, \ldots, n, \tag{7.2.7}$$

where a_i $(i = 1, 2, \ldots, n)$ may have any sign and there is no restriction on the coefficients b_{ij}. Our next problem is to determine the rate of entropy-production of the ecosystem governed by the system of ordinary differential equations (7.2.7). The non-trivial stationary state $(N_1^\star, N_2^\star, \ldots, N_n^\star)$ of the system (7.2.7) is given by

$$\sum_{j=1}^{n} b_{ij} N_j^\star = a_i, \; i = 1, 2, \ldots, n. \tag{7.2.8}$$

With the help of the relations (7.2.8), the system of Eq. (7.2.7) can be rewritten as:

$$\frac{\mathrm{d}}{\mathrm{d}t}(\ln N_i) = \sum_{j=1}^{n} b_{ij}(N_j^\star - N_j), \; i = 1, 2, \ldots, n. \tag{7.2.9}$$

The special form (7.2.9) of Lotka–Volterra equations allows for a productive thermodynamic formalism to be developed. Let us consider an association of n populations as a thermodynamic system open to flows of mass and energy and the number N_i of the individuals of each population, the thermodynamic state variables determining the system. For this, we assume that the non-equilibrium thermodynamic model just considered can be described completely by n generalized thermodynamic fluxes J_i and n generalized thermodynamic forces X_i (giving rise to these flows) which have the following functional relation with the basic thermodynamic variables N_i:

$$J_i = \frac{\mathrm{d}}{\mathrm{d}t}(\ln N_i) \text{ and } X_i = N_i^\star - N_i, \; i = 1, 2, \ldots, n. \tag{7.2.10}$$

Thus, the choice of J_i interpreted in ecological term is the specific rate of increase or decrease of the i-th population. The choice of X_i expresses the extent of deviation from the stationary state, and in that way, it gives a measure of the driving force that this population exercises on the system. With these choices of J_i and X_i the system of Eq. (7.2.9) can be reduced to the form:

$$J_i = \sum_{j=1}^{n} b_{ij} X_j, \; i = 1, 2, \ldots, n, \tag{7.2.11}$$

which are the linear phenomenological relations of non-equilibrium thermodynamics [9, 17]. The rate of entropy-production (evolutionary part) is then given by (Prigogine, 1967):

$$\frac{\mathrm{d}\overline{S}}{\mathrm{d}t} = \sum_{i=1}^{n} J_i X_i = \sum_{i=1}^{n} (N_i^\star - N_i) \frac{\mathrm{d}}{\mathrm{d}t}(\ln N_i). \tag{7.2.12}$$

Comparing this with the rate of entropy-production (7.2.6), we have

$$\frac{\Delta\mu_i}{T} = \frac{N_i - N_i^\star}{N_i}, \tag{7.2.13}$$

expressing the chemical potential difference $\Delta\mu_i$ in terms of the deviation of the non-stationary population from the stationary value. Feistel and Ebeling using ideal solution model of ecosystem obtained an analogous relation (Feistal and Ebeling 1990):

$$\Delta\mu_i = \mu_i - \mu_i^\star = -KT \ln\left(\frac{N_i^\star}{N_i}\right), \tag{7.2.14}$$

which reduces to the form (7.2.13) of the Lotka–Volterra system in the limiting case: $-1 < \dfrac{N_i^\star - N_i}{N_i} \ll 1.$

Let us now investigate the ecological significance of the change of entropy-production from the non-stationary state $(N_1, N_2, \ldots, N_n)$ to the stationary state $(N_1^\star, N_2^\star, \ldots, N_n^\star)$. Integrating (7.2.12) from $(N_1, N_2, \ldots, N_n)$ to $(N_1^\star, N_2^\star, \ldots, N_n^\star)$, we get

$$\Delta\overline{S} = \sum_{i=1}^{n}\left\{(N_i - N_i^\star) + N_i^\star \ln\left(\frac{N_i^\star}{N_i}\right)\right\}, \tag{7.2.15}$$

which is the measure of entropy-production in the transition from the non-equilibrium state $(N_1, N_2, \ldots, N_n)$ to the equilibrium state $(N_1^\star, N_2^\star, \ldots, N_n^\star)$ assuming that the system remains isothermal during the transition. Let us split up $(N10, N20, \ldots, Nn0)$ the entropy-production $\Delta\overline{S}$ into two parts:

$$\Delta\overline{S} = \Delta\overline{S}^{(1)} + \Delta\overline{S}^{(2)}, \text{ where } \Delta\overline{S}^{(1)} = \sum_{i=1}^{n}(N_i - N_i^\star);\ \Delta\overline{S}^{(2)} = \sum_{i=1}^{n} N_i^\star \ln\left(\frac{N_i^\star}{N_i}\right) \tag{7.2.16}$$

The first term $\Delta\overline{S}^{(1)}$ is the contribution due to change of the total population or energy and is zero when the total population remains fixed: $\sum_{i=1}^{n} N_i = \sum_{i=1}^{n} N_i^\star$. The second term $\Delta\overline{S}^{(2)}$ is of special significance for the ecosystem which may be called as the ecological entropy-production. It represents by its structure as the change of ecological diversity denoted by $-\Delta D$. Now, (7.2.15) can be rewritten as:

$$\Delta\overline{S} = \sum_{i=1}^{n} N_i^\star\left\{\left(\frac{N_i}{N_i^\star}\right) - \ln\left(\frac{N_i}{N_i^\star}\right) - 1\right\}. \tag{7.2.17}$$

We know that $z - \log(z) - 1 \geq 0$ for all $z > 0$ and the equality occurs only for $z = 1$. Therefore, $\Delta\overline{S} > 0$ if $N_i \neq N_i^\star$ for at least one i, and $\Delta\overline{S} = 0$ if $N_i = N_i^\star,\ \forall i$. Thus, there is still a positive contribution to entropy-production even if the total population does not change (provided $N_i \neq N_i^\star$ for at least one i). Therefore, we conclude that the energy is not only the factor which determines the evolution of the system, but there are some other factors such as the ecological diversity D which plays a significant role in the characterization of stability and evolution of the ecosystem (Chakrabarti et al. 1995).

7.3 Measure of Organization and Criteria of Evolution

We have the expression of entropy-production for the evolution part of the ecosystem. Let us now study the ecological significance of the thermodynamic criterion of stability based on this expression of entropy-production. We change to the state variables:

$$x_i = \ln\left(\frac{N_i}{N_i^\star}\right), \quad i = 1, 2, \ldots, n, \tag{7.3.1}$$

From (7.2.17) and (7.3.1), we have

$$\Delta\overline{S} = \sum_{i=1}^{n} N_i^\star(\mathrm{e}^{x_i} - x_i - 1) \Rightarrow \frac{\partial}{\partial x_i}(\Delta\overline{S}) = N_i^\star(\mathrm{e}^{x_i} - 1) = N_i - N_i^\star. \tag{7.3.2}$$

From (7.2.9) and (7.3.2):

$$\frac{\mathrm{d}x_i}{\mathrm{d}t} = -\sum_{j=1}^{n} b_{ij}\frac{\partial}{\partial x_j}(\Delta\overline{S}), \; i = 1, 2, \ldots, n. \tag{7.3.3}$$

The system of Eq. (7.3.3) indicates that $\Delta\overline{S}$ to be a *kinetic potential.* The time evolution of $\Delta\overline{S}$ can then be written in the following form (using (7.3.2) and (7.3.3)):

$$\frac{\mathrm{d}}{\mathrm{d}t}(\Delta\overline{S}) = -\sum_{i,j=1}^{n} b_{ij}\frac{\partial}{\partial x_i}(\Delta\overline{S})\frac{\partial}{\partial x_j}(\Delta\overline{S}), \tag{7.3.4}$$

which is positive or negative according to the negative or positive definiteness of the community matrix (b_{ij}). The *kinetic potential* or *entropy-production* $\Delta\overline{S}$ which is positive for all $N_i, N_i^\star > 0$ and is equal to zero for $N_i = N_i^\star$ $(i = 1, 2, \ldots, n)$ can be regarded as a Lyapunovthe functional. It is in fact very often chosen as the Lyapunov functional in the study of stability of various ecological and chemical reaction system [16, 20, 21]. The thermodynamic criteria of stability or instability of the system are then given by (Lavenda 1978):

$$\frac{\mathrm{d}}{\mathrm{d}t}(\Delta\overline{S}) \begin{cases} < 0 \Rightarrow \text{ stability,} \\ = 0 \Rightarrow \text{ marginal stability,} \\ > 0 \Rightarrow \text{ instability.} \end{cases} \tag{7.3.5}$$

In view of the Lyapunov function interpretation of the expression of entropy-production $\Delta\overline{S}$, the system of inequalities (7.3.5) serves as both dynamical and thermodynamic criteria of stability or instability of the ecosystem. Let us now investigate the ecological significance of the criteria of stability given by (7.3.5). We can write from (7.2.16) the entropy-production $\Delta\overline{S}$ as:

$$\Delta\overline{S} = \Delta N - \Delta D, \text{ where } \Delta N = \sum_{i=1}^{n}(N_i - N_i^\star) \text{ and } \Delta D = \sum_{i=1}^{n} N_i^\star \ln\left(\frac{N_i}{N_i^\star}\right), \tag{7.3.6}$$

where ΔD is the deviation from ecological diversity from its stationary value. The criteria of stability (7.3.5) then implies

$$\frac{dD}{dt} \geq \frac{dN}{dt} \tag{7.3.7}$$

which must be satisfied along a trajectory leading towards a non-trivial stable state. This means that if the community evolves towards a stable non-trivial stationary state, then the rate of change of diversity must be greater than or equal to the rate of biomass production. The reverse of the inequality in (7.3.7) leads to the onset of instabilities and the evolution of the community to a new regime resulting in the extinction of one or several species and thereby making the disintegration of the ecological community (Chakrabarti et al. 1995).

Self-organization or Evolution

The concept of self-organization or evolution can be explained from the view point of Prigogine's principle of *order through fluctuation* (Prigogine and Nicolis 1971). For this, we investigate the behaviour of a stationary state $(N_1^\star, N_2^\star, \ldots, N_n^\star)$ subject to external constraints (or environmental influences or disturbances). As a result of the disturbance, the stationary state $(N_1^\star, N_2^\star, \ldots, N_n^\star)$ is displaced to a neighbouring non-stationary state $(N_1, N_2, \ldots, N_n)$ such that

$$N_i = N_i^\star + \delta N_i, \; i = 1, 2, \ldots, n, \tag{7.3.8}$$

where δN_i $(i = 1, 2, \ldots, n)$ are the deviation or perturbation from the stationary values $(N_1^\star, N_2^\star, \ldots, N_n^\star)$. These deviations δN_i are a certain type of fluctuation (stochastic elements) as Prigogine emphasized to build a generalized thermodynamics (which includes a macroscopic theory of fluctuation).

We can define the extent of organization of a non-equilibrium state by the loss of evolutionary entropy $(\Delta\overline{S})$ in the transition from the non-equilibrium state $(N_1, N_2, \ldots, N_n)$ to the equilibrium state $(N_1^\star, N_2^\star, \ldots, N_n^\star)$ assuming that the system remains isothermal during the transition. Assuming δN_i $(i = 1, 2, \ldots, n)$ are very small and neglecting higher powers of δN_i, we have

$$\Delta\overline{S} = \sum_{i=1}^{n} N_i^\star \left\{ \left(\frac{N_i}{N_i^\star}\right) - \ln\left(\frac{N_i}{N_i^\star}\right) - 1 \right\}$$

$$= \sum_{i=1}^{n} N_i^\star \left\{ \left(\frac{N_i^\star + \delta N_i}{N_i^\star}\right) - \ln\left(\frac{N_i^\star + \delta N_i}{N_i^\star}\right) - 1 \right\} \tag{7.3.9}$$

$$= \sum_{i=1}^{n} \left\{ \delta N_i - N_i^\star \ln\left(1 + \frac{\delta N_i}{N_i^\star}\right)\right\} \approx \frac{1}{2}\sum_{i=1}^{n} \frac{(\delta N_i)^2}{N_i^\star} > 0.$$

The expression (7.3.9) is positive definite and can be considered as a Lyapunov function. The criteria of stability of the equilibrium state $(N_1^\star, N_2^\star, \ldots, N_n^\star)$ are then given by (using (7.3.5)):

$$\frac{\mathrm{d}}{\mathrm{d}t}(\Delta\overline{S}) = \frac{\mathrm{d}}{\mathrm{d}t}\left\{\frac{1}{2}\sum_{i=1}^{n}\frac{(\delta N_i)^2}{N_i^\star}\right\} < 0. \tag{7.3.10}$$

If this condition is fulfilled, then the deviation from the equilibrium state cannot grow in time so that the system will remain within the domain of attraction.

If the equilibrium state $(N_1^\star, N_2^\star, \ldots, N_n^\star)$ is stable, then the system will remain within the domain of attraction of the equilibrium state, and there can be no evolution of the system. If the external constraints are strong enough to drive the system sufficiently from the equilibrium state $(N_1^\star, N_2^\star, \ldots, N_n^\star)$, it will give rise to an ordered state. In that state, we must have the criteria of instability of the equilibrium state:

$$\frac{\mathrm{d}}{\mathrm{d}t}(\Delta\overline{S}) = \frac{\mathrm{d}}{\mathrm{d}t}\left\{\frac{1}{2}\sum_{i=1}^{n}\frac{(\delta N_i)^2}{N_i^\star}\right\} \geq 0, \tag{7.3.11}$$

hence the system will evolve out of the domain of attraction of the stationary state and so we have to wait until the system settles down to a new stationary state. Changed external constraints imposed on the system may result in another instability, and so on. Then a succession of instabilities arise from the external constraints (environmental influences) resulting in the increase of the ecological organization. The evolution of the ecosystem is thus associated with an irreversible increase in organization resulting in the transition of the system from an ordered state to a more ordered state, which is the Prigogine's principle of *order through fluctuation* (Prigogine and Nicolis 1971).

Probabilistic interpretation

Let us assume that the population densities $(N_1, N_2, \ldots, N_n)$ to be random variables in view of the many body aspect of the system and the influences of the fluctuating (random) environment on the system. In that case, the deviations $\delta N_i = N_i - N_i^\star$ are random. We assume $\delta N_i,\ i = 1, 2, \ldots, n$

to be small, and then the probability of small deviation from the stationary (equilibrium) state can be reduced to the Gaussian law of probability distribution due to Einstein (Prigogine and Nicolis 1971):

$$P = Pr(\delta N_1, \delta N_2, \cdots, \delta N_n) \propto \exp\left\{-\frac{\Delta \overline{S}}{k}\right\} = \exp\left\{-\frac{1}{2k}\sum_{i=1}^{n}\frac{(\delta N_i)^2}{N_i^\star}\right\}, \tag{7.3.12}$$

which is valid for large range of different stochastic multicomponent systems, where k is a positive constant and in thermodynamics it is Boltzmann's constant.

Let us analyse the significance of the probability of deviation stated in (7.3.12) for the stability of the stationary (equilibrium) state:

- Since the exponent in (7.3.12) is negative, the probability of deviations $\delta N_i = N_i - N_i^\star$ increases as we approach the stationary state or stated in other word, the stationary state $(N_1^\star, N_2^\star, \ldots, N_n^\star)$ is more probable than any other neighbouring state reached by a small deviations or fluctuations.
- This is, however, not sufficient to ensure stability of the equilibrium state. The condition (7.3.10) must be satisfied in order that the random deviation from the stationary state cannot grow in time.
- These two probabilistic criteria are equivalent to the criteria (7.3.9) and (7.3.10) derived from the consideration of Lyapunov function interpretation of entropy-production (or entropy change).

More explicit physical significance of (7.3.10)

The condition stated in (7.3.10) implies that

$$\sum_{i=1}^{n}\left(\frac{dN_i}{dt}\right)\delta N_i < 0 \Rightarrow \left(\frac{dN_i}{dt}\right)\delta N_i < 0, \quad i = 1, 2, \ldots, n, \tag{7.3.13}$$

which corresponds to the well-known generalization of *Le Chatelier-Braun principle* of chemical equilibrium. In context of ecology, it quantifies the ability of the ecosystem to react pliable to external factor (Svirezhev 2001). More explicitly, it implies that if a system in stationary state undergoes a variation δN_i in any factor characterizing the stationary state, the compensating change in the growth rate $\dot{N}_i$ must occur in opposite direction. This principle may then be adopted as an alternative way of testing the stability of a system (Svirezhev 2001; Chakrabarti and Ghosh 2009).

Another thermodynamic criteria of stability

Expanding the evolutionary entropy $(\overline{S})$ about the equilibrium state $(N_1^\star, N_2^\star, \ldots, N_n^\star)$, we have

$$\overline{S} \approx \overline{S}_{eq} + (\delta\overline{S}) + \frac{1}{2}(\delta^2\overline{S}). \tag{7.3.14}$$

Using (7.2.6), one obtains for the second variation of $(\overline{S})$ about the equilibrium state $(N_1^\star, N_2^\star, \ldots, N_n^\star)$:

$$\frac{1}{2}(\delta^2\overline{S}) = -\frac{1}{2T}\sum_{j=1}^{n}\left(\frac{\partial \mu_i}{\partial N_j}\right)_{\text{eq}} \delta N_i \delta N_j \leq 0. \tag{7.3.15}$$

From (7.2.12), the rate of entropy-production (evolutionary part) is given by

$$\sigma = \frac{\mathrm{d}\overline{S}}{\mathrm{d}t} = \sum_{i=1}^{n} J_i X_i. \tag{7.3.16}$$

In the case of chemical reactions (Prigogine and Nicolis 1971):

$$J_i = v_i, \text{ and } X_i = \frac{\mathcal{A}_i}{T} = -\frac{\sum_\rho \nu_{\rho i}\mu_\rho}{T} < 0, \tag{7.3.17}$$

where v_i is the reaction rate, $\mathcal{A}_i$ the corresponding affinity, T the temperature, μ_ρ the chemical potential of constituent ρ and $\nu_{\rho i}$ the stoichiometric coefficient of ρ in the i-th reaction.

Far from equilibrium, one obtains the following criteria (Glansdorff et al. 1974):

$$\left[\frac{\mathrm{d}}{\mathrm{d}t}\left(\frac{1}{2}\delta^2\overline{S}\right)\right]_{\text{along motion}} \begin{cases} > 0 \Rightarrow \text{ asymptotic stability,} \\ = 0 \Rightarrow \text{ marginal stability,} \\ < 0 \Rightarrow \text{ instability.} \end{cases} \tag{7.3.18}$$

From (7.3.16):

$$\delta\sigma = \delta_J\sigma + \delta_X\sigma, \text{ where } \delta_J\sigma = \sum_{i=1}^{n} X_i\delta J_i, \text{ and } \delta_X\sigma = \sum_{i=1}^{n} J_i\delta X_i. \tag{7.3.19}$$

For purely dissipative systems without convective motion:

$$\delta\sigma = \delta_X\sigma = \sum_{i=1}^{n} J_i\delta X_i = \sum_{i=1}^{n} \delta J_i\delta X_i \quad [\because J_i = (J_i)_{\text{eq}} + \delta J_i = \delta J_i]. \tag{7.3.20}$$

Referring to Eqs. (7.3.17) and (7.3.19) and assuming that the system is maintained at constant temperature, we have (at the steady state):

$$\sum_{i=1}^{n} v_i\delta\mathcal{A}_i = T\sum_{i=1}^{n} J_i\delta X_i \leq 0; \quad \sum_{i=1}^{n}(v_i)_{\text{eq}}\delta\mathcal{A}_i = T\sum_{i=1}^{n}(J_i)_{\text{eq}}\delta X_i = 0. \tag{7.3.21}$$

In (7.3.21), it is understood that i is such that the affinities $\mathcal{A}_i$ (or forces X_i) are all independent. This may always be achieved by forming suitable linear combinations of velocities (or fluxes) and affinities (or forces) of the individual reactions (populations). With this choice of v_i, $\mathcal{A}_i$ (or J_i, X_i), the equality in (7.3.21) implies that

$$(v_i)_{\text{eq}} = (J_i)_{\text{eq}} = 0, \quad \text{at the steady state.} \tag{7.3.22}$$

It is imagined that the steady state of the system is perturbed as a result of a random fluctuation or of an external disturbance, where δv_i, $\delta\mathcal{A}_i$ (or δJ_i, δX_i) be the fluctuations in v_i, $\mathcal{A}_i$ (or J_i, X_i):

$$\begin{aligned} v_i &= (v_i)_{\text{eq}} + \delta v_i = \delta v_i \Rightarrow J_i = (J_i)_{\text{eq}} + \delta J_i = \delta J_i, \\ \mathcal{A}_i &= (\mathcal{A}_i)_{\text{eq}} + \delta\mathcal{A}_i \Rightarrow X_i = (X_i)_{\text{eq}} + \delta X_i. \end{aligned} \tag{7.3.23}$$

The variation of entropy-production due to the fluctuations of the affinities (or forces) will be (cf. Eqs. (7.3.19) and (7.3.23)):

$$T\delta_X\sigma = T\sum_{i=1}^{n} J_i\delta X_i = T\sum_{i=1}^{n} \delta J_i\delta X_i = \sum_{i=1}^{n} v_i\delta\mathcal{A}_i = \sum_{i=1}^{n} \delta v_i\delta\mathcal{A}_i. \tag{7.3.24}$$

If $\delta_X\sigma < 0$, then according to inequality (7.3.21) its variation will also be negative and $\delta_X\sigma$ will never go to zero, and so the fluctuation will never be damped. We obtain therefore the stability criterion as:

$$T\delta_X\sigma = T\sum_{i=1}^{n} \delta J_i\delta X_i = \sum_{i=1}^{n} \delta v_i\delta\mathcal{A}_i \geq 0, \tag{7.3.25}$$

for all fluctuations or external disturbances compatible with the equations of evolution.

It can be proved that (Nicolis and Prigogine 1977):

$$\frac{\mathrm{d}}{\mathrm{d}t}\left(\frac{1}{2}\delta^2\overline{S}\right) = \delta_X\sigma = \sum_{i=1}^{n} \delta J_i\delta X_i. \tag{7.3.26}$$

Here, $\delta_X\sigma$ is the excess entropy-production due to fluctuations or external disturbances. Inequalities (7.3.15) and (7.3.25) constitute an equivalent stability criterion. $\delta^2\overline{S}$ is seen to play the role of a thermodynamic Lyapounov function by virtue of (7.3.15), (7.3.25) and (7.3.26).

References

Chakrabarti, C.G., Ghosh, K.: Non-equilibrium thermodynamics of ecosystems: entropic analysis of stability and diversity. Ecol. Model. **220**, 1950–1956 (2009)

Chakrabarti, C.G., Ghosh, S., Bhadra, S.: Non-equilibrium thermodynamics of Lotka-Volterra ecosystems: stability and evolution. J. Biol. Phys. **21**, 273–284 (1995)

Feistal, R., Ebeling, W.: Evolution of Complex Systems. Kluwer Academic Publishers, Dordrecht (1990)

Glansdorff, P., Nicolis, G., Prigogine, I.: The thermodynamic stability theory of non-equilibrium states. Proc. Nat. Acad. Sci. USA **71**(1), 197–199 (1974)

Lavenda, B.H.: Thermodynamics of Irreversible Processes. MacMillan, London (1978)

Nicolis, G., Prigogine, I.: Self-Organization in Non-Equilibrium Systems. Wiley Interscience, New York (1977)

Prigogine, I.: Introduction to Thermodynamics of Irreversible Processes. Interscience Publishers, New York (1967)

Prigogine, I., Nicolis, G.: Biological order, structure and instability. Quant. Rev. Biophys. **4**, 107–148 (1971)

Svirezhev, Y.M.: In: Jorgensen, S.E. (ed.) Thermodynamics and Ecological Modelling. Lewis Publishers, New York (2001)

Svirezhev, Yu.M., Logofet, D.O.: Stability of Biological Communities. MIR, Moscow (1978)

Chapter 8
Stability of a Social Group

8.1 Introduction

In this chapter, we have considered a system of social group which satisfies the following postulates (Simon 1967):

1. Interaction intensity is due to the combined effects of two communication causes: friendliness and activity.
2. If a group characterized by initial friendliness is induced to interact, friendship will increase, but there will be a lag.
3. Activity rate accommodates itself to the level of friendliness and to the activity imposed externally.

Here, we have discussed the stability of equilibrium of this system by means of loop analysis based on Levins's formulation (Levins 1975) in which the interactions between species can be specified in a qualitative but not a quantitative way. The present chapter also consists of the thermodynamic and stochastic modelling of the system and study of the stability of equilibrium from the consideration of thermodynamic and stochastic criteria of stability.

8.2 Mathematical Equations: Stability of Equilibrium State

Let $I(t)$, $F(t)$, $A(t)$ and $E(t)$ denote the intensity of interaction, the level of friendliness among the members, the amount of activity carried on by members of the group and the amount of activity imposed by external influences respectively. We consider a social group (a group of persons) whose behaviour can be characterized by these four variables, all functions of time t. Then the mathematical model of this system of social group developed by Simon (1967) is

© Springer Nature Singapore Pte Ltd. 2021

G. Samanta, *Deterministic, Stochastic and Thermodynamic Modelling of some Interacting Species*, Forum for Interdisciplinary Mathematics,
https://doi.org/10.1007/978-981-16-6312-3_8

$$
\begin{aligned}
I &= a_1 F + a_2 A \\
\frac{\mathrm{d}F}{\mathrm{d}t} &= b(I - \beta F) \\
\frac{\mathrm{d}A}{\mathrm{d}t} &= c_1(F - \gamma A) + c_2(E - A),
\end{aligned}
\tag{8.2.1}
$$

where $a_1, a_2, b, \beta, c_1, c_2$ and γ are coefficients assumed to be positive constants. Here, $I(t)$ and $A(t)$ are endogenous (dependent) variables whose values are determined within the system, while $E(t)$ is an exogenous (independent) variable.

From system (8.2.1), we have

$$
\begin{aligned}
\frac{\mathrm{d}F}{\mathrm{d}t} &= a_{11} F + a_{12} A, \\
\frac{\mathrm{d}A}{\mathrm{d}t} &= a_{21} F + a_{22} A + c_2 E,
\end{aligned}
\tag{8.2.2}
$$

where $a_{11} = -b(\beta - a_1)$, $a_{12} = b a_2$, $a_{21} = c_1$ and $a_{22} = -(c_1\gamma + c_2)$.

Here, we assume that E is a positive constant. The system (8.2.2) has stationary (equilibrium) solution:

$$
\begin{aligned}
F_e &= \frac{a_2 c_2 E}{(\beta - a_1)(c_1\gamma + c_2) - a_2 c_1}, \\
A_e &= \frac{(\beta - a_1) c_2 E}{(\beta - a_1)(c_1\gamma + c_2) - a_2 c_1}, \\
&\text{provided } \beta > a_1 + \frac{a_2 c_1}{(c_1\gamma + c_2)}.
\end{aligned}
\tag{8.2.3}
$$

Let $F = F_e + x_1$ and $A = A_e + x_2$, then system (8.2.2) gives

$$
\begin{aligned}
\frac{\mathrm{d}x_1}{\mathrm{d}t} &= a_{11} x_1 + a_{12} x_2, \\
\frac{\mathrm{d}x_2}{\mathrm{d}t} &= a_{21} x_1 + a_{22} x_2.
\end{aligned}
\tag{8.2.4}
$$

The eigenvalue equation for the system (8.2.4) is

$$
\lambda^2 - (a_{11} + a_{22})\lambda + (a_{11}a_{22} - a_{12}a_{21}) = 0. \tag{8.2.5}
$$

Therefore, the eigenvalues are

$$\lambda_1, \lambda_2 = \frac{-\Lambda \pm \sqrt{\Delta}}{2},$$

where $\Lambda = -(a_{11} + a_{22}) = b(\beta - a_1) + (c_1\gamma + c_2) > 0,$ by (1.3) and

$$\begin{aligned}\Delta &= (a_{11} + a_{22})^2 - 4(a_{11}a_{22} - a_{12}a_{21}) \\ &= \{b(\beta - a_1) + (c_1\gamma + c_2)\}^2 - 4b(c_1\gamma + c_2)\left\{\beta - \left(a_1 + \frac{a_2c_1}{c_1\gamma + c_2}\right)\right\} < \Lambda^2 \\ &\Rightarrow \lambda_1, \lambda_2 \text{ have negative real parts.}\end{aligned} \tag{8.2.6}$$

Therefore, if the stationary state of the system (8.2.1) exists, then it is asymptotically stable. Now (8.2.3) implies

$$\beta F_e > a_1 F_e. \tag{8.2.7}$$

What this means is that the quantity of interaction (βF_e) necessary to produce the stationary level of friendliness must be greater than the amount of communication ($a_1 F_e$) resulting from this equilibrium degree of friendliness, if equilibrium (stationary) state exists which is also asymptotically stable. This is a valuable conclusion derived mathematically and easily tested empirically (Mapes 1971).

If E is a negative constant, then the system (8.2.2) has equilibrium solution (8.2.3), provided

$$a_1 < \beta < \frac{a_2c_1}{(c_1\gamma + c_2)}. \tag{8.2.8}$$

In this case, one of the eigenvalues λ_1, λ_2 is positive and so they are of opposite in sign. Hence, the equilibrium solution (8.2.3) is unstable which is a saddle point.

8.3 Stability of Equilibrium State: Method of Loop Analysis

Let $X_1 = F$ and $X_2 = A$, then the system (8.2.2) can be represented by a diagram based on Levins's formulation (Levins 1975), in which each variable is represented by a point (or a vertex, or a circle) and the relations among variables appear as oriented links connecting the vertices so that the line connecting X_j to X_i represents the interaction or effects of X_j on X_i and corresponds to the element a_{ij}. A link or series of links that leaves and eventually reenters the same vertex is called a *loop*. Corresponding to the diagonal matrix elements a_{ii}, there are loops that connect each X_i to itself, termed *self-loops*. When two loops share no vertices in common, we

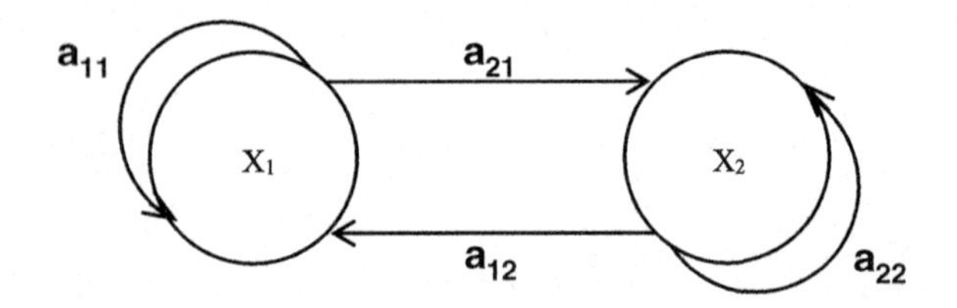

Fig. 8.1 Pictorial representation of system (8.2.2) based on Levins's formulation

refer to the loops as *disjunct*, otherwise these are called *conjunct*. The *feedback* at level k in a system is defined by

$$R_k = \sum_m (-1)^{m+1} L(m,k), \tag{8.3.1}$$

where $L(m,k)$ is the product of k vertices (i.e. coefficients a_{ij}) which form m disjunct loops and the summation is taken over all m and all possible products involving m loops.

The system (8.2.2) will be stable, i.e. it returns to its equilibrium state after a perturbation, if and only if (Levins 1975):

$$R_k = \sum_m (-1)^{m+1} L(m,k) < 0, \quad k = 1, 2 \quad (\text{provided } E > 0). \tag{8.3.2}$$

Here $a_{11} = -b(\beta - a_1) < 0, a_{12} = ba_2 > 0, a_{21} = c_1 > 0$ and $a_{22} = -(c_1\gamma + c_2) < 0$.

Fig. 8.1 represents this case. Here, friendliness promotes activity rate (since $a_{21} > 0$) and is symbolized by an arrow from X_1 to X_2 and conversely activity rate promotes friendliness (since $a_{12} > 0$) and is symbolized by an arrow from X_2 to X_1. The growths of friendliness and activity rate are self-damped (since $a_{11}, a_{22} < 0$) and are expressed by the self-loops with circles on X_1 and X_2, respectively. Now,

$$F_1 = \sum_i a_{ii} = -\{b(\beta - a_1) + (c_1\gamma + c_2)\} < 0,$$

$$\begin{aligned} F_2 &= \sum_{i<j} a_{ij}a_{ji} - \sum_{i<j} a_{ii}a_{jj} = a_{12}a_{21} - a_{11}a_{22} \\ &= -b(c_1\gamma + c_2)\left\{\beta - \left(a_1 + \frac{a_2c_1}{c_1\gamma + c_2}\right)\right\} < 0 \end{aligned} \tag{8.3.3}$$

$$\left(\text{since} \quad \beta > a_1 + \frac{a_2c_1}{c_1\gamma + c_2}, \quad \text{for existence of equilibrium state}\right).$$

Therefore, the system is stable.

This section presents a new method of analysis of stability behaviour of a system. If the interaction of a system of social group can be specified in a qualitative but not a quantitative way, then this formalism proves particularly useful for examining the properties of social groups.

8.4 Thermodynamic Model and Stability

The social group system can be considered as a network of flows of energy and biomass. An appropriate modelling of such a system can be carried out by the thermodynamics of irreversible processes. To develop a thermodynamic model of the system governed by (8.2.4), we have to choose the thermodynamic fluxes and forces properly. We choose the thermodynamic fluxes J_i and forces $\mathcal{X}_i$ as:

$$J_i = \frac{\mathrm{d}x_i}{\mathrm{d}t}, \quad \mathcal{X}_i = -x_i, \quad i = 1, 2. \tag{8.4.1}$$

The choice (8.4.1) is appropriate in the sense that in the equilibrium state (F_e, A_e): the thermodynamic forces $\mathcal{X}_i$, which are the deviation from the equilibrium state (F_e, A_e), vanish and correspondingly the fluxes and forces also stop. With these choices, we can thus rewrite the system (8.2.4) as:

$$\begin{aligned} J_1 &= -a_{11}\mathcal{X}_1 - a_{12}\mathcal{X}_2, \\ J_2 &= -a_{21}\mathcal{X}_1 - a_{22}\mathcal{X}_2, \end{aligned} \tag{8.4.2}$$

which are the linear phenomenological relations of irreversible thermodynamics. If S is the entropy of the system for the non-equilibrium state near the equilibrium state, then it can be expanded about the equilibrium state and we can write:

$$S \approx S_{\text{eq}} + (\delta S) + \frac{1}{2}(\delta^2 S). \tag{8.4.3}$$

Then the thermodynamic criteria of stability of the equilibrium state are given by (see Chap. 7):

$$\frac{\mathrm{d}}{\mathrm{d}t}\left(\frac{1}{2}\delta^2 S\right) = \sum_i \delta J_i \delta \mathcal{X}_i > 0. \tag{8.4.4}$$

For the system (8.4.2), this becomes

$$a_{11}(\delta\mathcal{X}_1)^2 + (a_{12} + a_{21})\delta\mathcal{X}_1\delta\mathcal{X}_2 + a_{22}(\delta\mathcal{X}_2)^2 < 0. \tag{8.4.5}$$

It implies negative definiteness of the matrix (a_{ij}). So, the criteria of thermodynamic stability of the equilibrium state become

$$a_{11} < 0, a_{22} < 0, 4a_{11}a_{22} > (a_{11} + a_{22})^2. \tag{8.4.6}$$

First two conditions are satisfied ($\because\ \beta > a_1 + \dfrac{a_2 c_1}{c_1\gamma + c_2}$, for existence of equilibrium state). The third condition gives

$$\beta > a_1 + \frac{(ba_2 + c_1)^2}{4b(c_1\gamma + c_2)}. \tag{8.4.7}$$

Now,

$$a_2 c_1 - \frac{(ba_2 + c_1)^2}{4b} = -\frac{(ba_2 - c_1)^2}{4b} \leq 0. \tag{8.4.8}$$

Therefore, the thermodynamic criteria of stability imply the existence and stability (deterministic) of the equilibrium state but the converse is not true in general. If $c_1 = a_2 b_1$, i.e. $\dfrac{\partial}{\partial A}\left(\dfrac{\mathrm{d}F}{\mathrm{d}t}\right) = \dfrac{\partial}{\partial F}\left(\dfrac{\mathrm{d}A}{\mathrm{d}t}\right)$, i.e. if the rate of change with respect to activity of the rate of change of friendliness with respect to time is equal to the rate of change with respect to friendliness of the rate of change of activity with respect to time, then the existence of equilibrium state implies thermodynamic stability and deterministic stability of this system.

8.5 Stochastic Model and Stability

The stochastic extension of the system of equations (8.2.2) is given by the following system of stochastic differential equations:

$$\begin{aligned} \frac{\mathrm{d}F}{\mathrm{d}t} &= a_{11}F + a_{12}A + \Gamma_1(t), \\ \frac{\mathrm{d}A}{\mathrm{d}t} &= a_{21}F + a_{22}A + c_2 E + \Gamma_2(t), \end{aligned} \tag{8.5.1}$$

where $\Gamma_i(t),\ i = 1, 2$ are the random perturbation terms due to the overall effect of the internal fluctuations between the level of friendliness among the members and the amount of activity carried on by members of the group. These random perturbation terms $\Gamma_i(t)$ satisfy the following properties:

$$\langle \Gamma_i(t) \rangle = 0,\ \langle \Gamma_i(t)\Gamma_j(t') \rangle = D_{ij}\delta(t - t'),\ D_{ij} = D\delta_{ij},\ i, j = 1, 2, \tag{8.5.2}$$

where $\langle \cdot \rangle$ represents the average over the ensemble of the stochastic process, D is the intensity of the noises, δ_{ij} is the *Kronecker delta* and $\delta(t)$ denotes the *Dirac delta* function.

The fluctuation intensities (variances) in F and A at any arbitrary instant t satisfying the system of stochastic differential equations (8.5.1) are given by (Risken 1984):

$$\sigma_F^2(t) = \int_0^t \left\{ \sum_{k=1}^{2} \sum_{s=1}^{2} G_{1k}(t')G_{1s}(t')D_{ks} \right\} \mathrm{d}t',$$
$$\sigma_A^2(t) = \int_0^t \left\{ \sum_{k=1}^{2} \sum_{s=1}^{2} G_{2k}(t')G_{2s}(t')D_{ks} \right\} dt', \tag{8.5.3}$$

where $G_{ij} - \sum_{k=1}^{2} a_{ik} G_{kj} = 0,\ i, j = 1, 2.$

After some simplifications, we have:

$$\begin{aligned}
\sigma_F^2(t) &= \frac{D}{2} \left\{ \frac{1}{\lambda_1}(c_1'^2 + c_2'^2)(\mathrm{e}^{2\lambda_1 t} - 1) + \frac{1}{\lambda_2}(c_1''^2 + c_2''^2)(\mathrm{e}^{2\lambda_2 t} - 1) \right. \\
&\quad \left. + \frac{4}{\lambda_1 + \lambda_2}(c_1'c_1'' + c_2'c_2'')\{\mathrm{e}^{(\lambda_1+\lambda_2)t} - 1\} \right\}, \\
\sigma_A^2(t) &= \frac{D}{2} \left\{ \frac{1}{\lambda_1}(c_3'^2 + c_4'^2)(\mathrm{e}^{2\lambda_1 t} - 1) + \frac{1}{\lambda_2}(c_3''^2 + c_4''^2)(\mathrm{e}^{2\lambda_2 t} - 1) \right. \\
&\quad \left. + \frac{4}{\lambda_1 + \lambda_2}(c_3'c_3'' + c_4'c_4'')\{\mathrm{e}^{(\lambda_1+\lambda_2)t} - 1\} \right\},
\end{aligned} \tag{8.5.4}$$

where c_i', c_i'' are arbitrary constants and λ_i are given by (8.2.6). Therefore, $\sigma_F^2(t)$, $\sigma_A^2(t)$ converge exponentially to the limiting variances:

$$\begin{aligned}
\sigma_F^2(\infty) &= \frac{D}{2} \left\{ \left(\frac{|\, c_1' \,|}{\sqrt{n_1}} - \frac{|\, c_1'' \,|}{\sqrt{n_2}} \right)^2 + \left(\frac{|\, c_2' \,|}{\sqrt{n_1}} - \frac{|\, c_2'' \,|}{\sqrt{n_2}} \right)^2 \right. \\
&\quad \left. + 2\frac{(\sqrt{n_1} \pm \sqrt{n_2})^2}{(n_1 + n_2)\sqrt{n_1 n_2}}(|\, c_1'c_1'' \,| + |\, c_2'c_2'' \,|) \right\}, \\
\sigma_A^2(\infty) &= \frac{D}{2} \left\{ \left(\frac{|\, c_3' \,|}{\sqrt{n_1}} - \frac{|\, c_3'' \,|}{\sqrt{n_2}} \right)^2 + \left(\frac{|\, c_4' \,|}{\sqrt{n_1}} - \frac{|\, c_4'' \,|}{\sqrt{n_2}} \right)^2 \right. \\
&\quad \left. + 2\frac{(\sqrt{n_1} \pm \sqrt{n_2})^2}{(n_1 + n_2)\sqrt{n_1 n_2}}(|\, c_3'c_3'' \,| + |\, c_4'c_4'' \,|) \right\},
\end{aligned} \tag{8.5.5}$$

where $n_i = -\lambda_i$.

Now, in an unvarying environment, the eigenvalues of the 2×2 interaction matrix of the system (8.2.4) are $\lambda_1, \lambda_2 = \frac{-\Lambda \pm \sqrt{\Delta}}{2}$, by (8.2.6). Therefore, the stability determining quantity is Λ and the deterministic stability criterion is satisfied, since $\Lambda > 0$. In a stochastic environment, whose random fluctuations have variance D, the stability provided by the interaction dynamics is again characterized by Λ. It is no longer enough that $\Lambda > 0$, for if $D \gg \Lambda$, the system exhibits large fluctuations, which rapidly lead to extinction. In the intermediate region where Λ and D are commensurate, the system is likely to undergo significant fluctuations, even though they persist for long times. If $D \ll \Lambda$, the fluctuations become very small; therefore, the deterministic solution is recovered.

8.6 Discussion

In this chapter, we have studied the stability behaviour of the equilibrium state (provided it exists) of a social group developed by Simon (1967) by means of loop analysis based on Levins's formulation and from the consideration of thermodynamic and stochastic criteria of stability. We have seen that the existence of equilibrium state ensures the asymptotic stability of equilibrium state, whereas the thermodynamic model needs different criteria for stability. The thermodynamic criteria of stability ensure the existence and stability (deterministic) of equilibrium state but the converse is not true in general. The stochastic model not only depends on the parameters of the system but also on the fluctuation.

References

Levins, R.: Evolution in Communities Near Equilibrium. In: Cody, M.L., Diamond, J.M. (eds.) Ecology and Evolution of Communities. The Belknap Press of Harverd University Press, Cambridge (1975)
Mapes, R.: Mathematics and Sociology. B.T. Batsfor Ltd., London (1971)
Risken, H.: The Fokker-Planck Equation. Springer, Berlin (1984)
Simon, H.: Models of Man. Wiley, New York (1967)

Printed in Great Britain
by Amazon